TOSHIBA MEDIUM PLC PRIMER

TOSHIBA MEDIUM PLC PRIMER

Ed Dropka

Butterworth-Heinemann
Boston Oxford Melbourne Singapore
Toronto Munich New Delhi Tokyo

∞ Recognizing the importance of preserving what has been written, Butterworth–Heinemann prints its books on acid-free paper whenever possible.

Library of Congress Cataloging-in-Publication Data

Dropka, Ed, 1937–
Toshiba medium PLC primer / by Ed Dropka.
p. cm.
Includes index.
ISBN 0-7506-9694-X (acid-free paper)
1. Programmable logic devices. 2. Programmable controllers.
I. Title.
TK7872.L64D76 1995
629.8′9—dc20 95-19003
CIP

British Library Cataloguing-in-Publication Data

The catalogue record for this book is available from the British Library.

The publisher offers discounts on bulk orders of this book. For information, please write:
Manager of Special Sales
Butterworth–Heinemann
313 Washington Street
Newton, MA 02158–1626

10 9 8 7 6 5 4 3 2 1

Printed in the United States of America

CONTENTS

TABLE OF CIRCUITS

PREFACE

The merger of Programmable Logic Controllers (PLCs) and Relay Ladder Logic (RLL) has created a powerful combination of both old and new technologies. The RLL language was first used with electromechanical relay systems that controlled the operation of various types of industrial machines. The PLC itself was initially designed as a "relay replacer" for these machines. As a result, PLCs were developed with RLL as the operating language.

PLCs have now been developed to the point where they can perform a much wider range of operations than simple relay replacement. Today PLCs are being used in control applications in a wide variety of manufacturing plants and processing plants all around the world.

There are many brands of PLCs in the marketplace today, and nearly all of them support the RLL language. I have chosen to use the Toshiba brand of PLCs and the Toshiba EX-PDD250 software package for personal computers (PCs) as the basis for the *Toshiba Medium PLC Primer.*

The comments in this book about RLL are generally valid for any brand of PLC that uses RLL as the operating language. Also included are thirty-one circuits, along with a written description of each circuit. These circuits should help the reader become more familiar with PLCs and RLL.

The reader should be aware that other brands of PLCs use slightly different methods of programming the various RLL devices involved. I have not attempted to include these variations for all PLC brands. In spite of this, the *Toshiba Medium PLC Primer* can still provide great benefits to the reader who is interested in working with another brand of PLC. To work with another brand, simply replace the Toshiba device with the same device of the other brand in the circuit as you copy the circuit to your RLL software program in your PC.

* * * * * *

1 PLCs AND RELAY LADDER LOGIC

AN INTRODUCTION TO PLCs

Welcome to the world of PLCs!

And welcome to *The PLC Primer: An Introduction to Medium Programmable Logic Controllers and Relay Ladder Logic for Toshiba Medium PLCs,* also known as the *Toshiba Medium PLC Primer.*

Some of you may be asking, "Just what is a PLC?"

PLC stands for programmable logic controller. PLCs are electronic control devices that are used in a variety of industries, from manufacturing plants to processing plants. A few examples of operations in which PLCs are used include automotive assembly plants, automotive parts manufacturing plants, mineral processing plants, mines, oil and gas refineries, petroleum processing plants, semiconductor manufacturing plants, steel mills, water treatment plants, and waste water treatment plants.

A basic PLC has three components—the central processing unit (CPU), inputs, and outputs. The *inputs* and the *outputs* are commonly known as the I/O and include the terminal strips where field devices are connected. The field devices that serve as inputs come in a wide variety. The following are some types of input devices used:

Float switches
Flow switches
Level switches
Limit switches
Maintained push buttons
Momentary push buttons
Motion sensors
Photo-eye sensors
Pressure switches
Selector switches
Temperature switches
Thumbwheel switches

The field devices that serve as outputs also come in a wide variety of devices. The following are some types of output devices used:

AC variable frequency drives	Pilot lights
Alarms (all types)	Pumps
Fans	Relays (large)
Heaters	Solenoids
Motors (small)	Solid state starters
Motor starters	Valves

The CPU contains a microprocessor, which means that a PLC is basically a specialized computer that has been designed to control the operation of machines and processes within the harsh environment of a plant or operation. The language used to program the PLC to perform the logic required to connect the field inputs to the field outputs is called *relay ladder logic* (RLL). The RLL language is programmed by means of special software with a personal computer (PC). The program is then transferred to the CPU of the PLC to control a specific operation.

An example of an input is a START push button, and an example of an output is a motor. A simple PLC circuit uses the RLL language to have the START push button run the motor as long as the push button is held down. A more complex circuit has the START push button start the motor only after a time delay has elapsed and after a series of related safety sequences has been executed.

The interesting thing about PLCs is that once a field input or a field output has been physically wired to the PLC, the circuit can be revised through the RLL software on a PC without making any wiring changes to the PLC. This is true regardless of how simple the circuit is at the beginning or how complex it becomes.

AN INTRODUCTION TO TOSHIBA PLCs

Welcome to the world of Toshiba Medium PLCs!

This book is a joint effort between the author and the PLC Marketing and Applications Department of Toshiba International. It is called a *primer* because it is designed for electricians, engineers, students, and technicians who are not familiar with PLCs and RLL.

My intention is to bring you up to half-speed. As you work your way through this book and the EX-PDD software, you will see that I have not touched on all of the features of EX-PDD. The goal has been to cover only the features that are the most necessary to accelerate your learning. By the time you complete this book, you will have enough information at your disposal to get up to full-speed on your own, based on the experience you will gain through practice.

You can copy the circuits in this book directly into your programs. The spacing between the rungs in the ladder diagrams has been increased to make the circuits easier to read than those produced on the screen by the Toshiba EX-PDD software. The book also shows you how to test the programs.

This book covers all Toshiba Medium PLCs. That includes the M20 and M40, the EX100, the EX250, and the EX500. A second book covers the Toshiba Small PLCs, including the EX14B, the EX28B, the EX20, the EX20+, the EX40, the EX40+, and the EX40H. The comments in this book about RLL are generally valid for both the medium and small series of Toshiba PLCs.

The Toshiba Medium PLC Primer has been written primarily for use with the Toshiba EX-PDD software and should be used in conjunction with your Toshiba EX-PDD manual and the manual for your particular PLC. Keep these manuals at your side; they will come in handy, and we don't intend to reprint all of that information in this book. Using the EX-PDD software on your PC or laptop computer makes the job of writing and debugging a circuit much easier.

As you begin to work with the PLC manual, you will notice that it was not written for use with the EX-PDD software package. The PLC manual was specifically written for use with the Toshiba LCD Programmer. If you are programming your PLC with a Toshiba handheld programmer, you will still be able to make use of most of the material in this book.

Figures 1.1 through 1.3 show some of these Toshiba Medium PLC models. Figure 1.1 shows one of the M40 PLCs. The M20 PLC looks the same but is 3.15 inches shorter than the M40.

Figure 1.3 shows one of the EX250 PLCs. The EX500 PLC looks the same as the EX250. In fact, both PLCs use the same I/O modules.

THE M20 AND M40 PLCs

Toshiba has two PLCs in this medium series of PLCs that are brick-style—the M20 and the M40. Figure 1.4 shows the M20 PLC with an expansion rack.

These two models are essentially hybrid PLCs. They are called *hybrid* because of the following features:

- They have the same footprint size as the small series of PLCs, which includes the EX20, the EX20+, the EX40, the EX40+, and the EX40H.
- They contain the CPU of the medium series of PLCs, which includes the EX100, the EX250, and the EX500.

Figure 1.1 M40 PLC

Figure 1.2 EX100 PLC

Figure 1.3 EX250 PLC

Figure 1.4 M20 with Expansion Rack

- The M20 and the M40 can be expanded with one expansion unit. This unit can be either an M20 or an M40 expansion module, or it can be a two-slot or a four-slot expansion rack from the M20/M40 series.
- The standard EX100 series modules can be installed in these M20/M40 two-slot and four-slot expansion racks. This includes all of the EX100 modules, the high-density I/O modules, the analog I/O modules, the axis position modules, and even the ASCII/BASIC modules.
- The EX-PDD programming software used is the EX-PDD250 version, which is used with the medium series of PLCs. It has a substantially greater number of programming functions and capabilities than does the EX-PDD2040, which is used with the small series of PLCs.

WHAT IS RELAY LADDER LOGIC?

You are probably saying to yourself, "That's a good question!"

RLL is *not* a new programming language. In fact, RLL was one of the first graphical programming languages. Its roots go back more than seventy years, when relays were coming into prominence in the telephone industry. However, RLL seemed to rise to the occasion and become the dominant method of electrical control of equipment during the early growth periods in the machine tool and automatic assembly machine industries.

As relays became important in the control of machine sequences, RLL became the language of control and the means of communication between the circuit designer and the electrician. The electricians needed the RLL diagrams not only for wiring the machine during fabrication but also to keep the machines operating once they were in production. The key to troubleshooting a machine built by a company many miles away was the ability to read and understand the RLL diagram for the machine.

The word *ladder* within the term *RLL* is the one that is probably the most confusing. Yet it is the least technical term involved. Figure 1.5 shows a basic RLL diagram.

If you were to strip away all of the devices that operate within the RLL, only the basic structure would be left standing, and it would look like a common ladder. The horizontal lines are called *rungs*. The *uprights* of the ladder represent the voltage source, the left upright being the *hot* side, and the right upright being the *common* side. Therefore, current flow is represented as being from the left upright to the right upright.

Before the advent of PLCs and PCs, all RLL diagrams were drawn on a drafting board by a circuit designer. The diagram contained all of the electrical devices needed to wire the machine, including all wire num-

Figure 1.5 Basic RLL Diagram

```
120 VAC                                                          COMMON

|                                                                     |
|-| |----|/|-------------------------------------------------( )-|
|X000    X001                                                R000     |
|                                                                     |
|                                                                     |
|-| |----|/|-------------------------------------------------( )-|
|R000    R004                                                Y020     |
|                                                                     |
|                                                                     |
|-| |----|/|-------------------------------------------------( )-|
|R000    R006                                                Y021     |

                        CURRENT FLOW ----->
```

bers. Of course there were more wires in those days, because all relays were of the electromechanical type. Therefore, each coil and every contact on the relay required two separate wires.

The introduction of the PLC eliminated many of the electromechanical relays previously required, which has greatly reduced the physical size of the control panels used. The PLC, then, has also greatly reduced the amount of the wiring required to connect the relays to each other and to the rest of the electrical equipment. In addition to the foregoing benefits, PLC systems provide greater reliability because of the reduced failure rate of PLCs compared with electromechanical relays.

Your next question probably is: "But many PLCs actually use small electromechanical relays as their outputs. If that is true, how can PLCs provide all the benefits you mentioned?"

Although the final output relays of a PLC may, in some models, be an electromechanical relay, those output relays represent only a small portion of the relays that the PLC has been able to convert from external electromechanical relays into internal transistors (on a semiconductor chip) that function as the internally generated relays. That is where the real benefits originate.

These internally generated PLC relays perform the control functions previously performed by hard-wired relays. They interface with the remaining hard-wired portion of the system through their own outputs, which may be electromechanical relays. The result is that an M40 PLC, with sixteen outputs, can replace the following:

960 relays

512 link relays

128 timers

96 counters

The 960 relays also can be used to create flip/flops, drum (or step) sequencers, and shift registers.

Can you imagine the size of the control panel required to mount all of those devices? Even if you were to use ice-cube style relays, the panel would be substantially larger than an M40 PLC, which measures only 4.9 inches wide by 3.1 inches high by 12.6 inches long.

Because these devices are now internal to the PLC, no hard-wiring is required between them. This means that changes to the RLL circuit required during system debugging or during subsequent system changes can be made within the PLC, without hard-wire modifications.

DESIGNING RELAY LADDER LOGIC CIRCUITS

To begin with, the job of designing RLL circuits can be an extremely interesting, challenging, and rewarding task. However, it does require patience, perseverance, and practice. The best way to begin work on a circuit is not with the RLL itself but with a detailed written description of how the completed circuit will operate.

The written description can be in the form of a timing diagram, a sequence chart, or an operational description. It is best if the end-user writes the description. Who actually writes the description, however, is not nearly as important as the fact that it needs to be completed before the RLL is started.

The word-processing program in your PC is the best way to work on the written description. If a word processor is not available, write the description by hand—it is extremely important that it be done.

The second step is equally important. The written description must be reviewed with the people who are the most concerned about the way the circuit will function once the equipment is in operation. This may be the engineering manager, the maintenance manager, the operational manager, or a combination of all of them.

The review is an extremely valuable step in the circuit design process. Changes invariably are made in the written description, regardless of how much discussion went into the first draft. Committing a design scheme to writing has a way of generating many valid and previously unasked questions. During this discussion, be sure that these individuals not only agree with what you have written, but that they, as the users, understand it. The more complete the discussion at this point, the better will be the results at the end of the project.

When you have a revised and final document that represents the complete operation, bear in mind that you can still look forward to changes. Why? Because the end result of the way your circuit operates may still not be the way it was visualized by all the people involved in the written description.

As far as the actual design of the circuit is concerned, it is best to break up the operation into as many small segments as possible. Don't try to design more than you can handle at any given time. Interruptions to your thought process and your circuit-design efforts can be disastrous. Designing small segments of circuits in each work session minimizes the effect of interruptions.

When you start your design, you can work out some of the circuits on your PC screen with the EX-PDD software. However, you will quickly begin to design circuits that are longer than one page or are linked to other circuits within the program. At that time, it is advisable to return to the good old pencil-and-paper method to sketch the circuit. When that portion of the circuit has been completed, transfer it to the EX-PDD program in your PC.

Transfer all your small circuits, one at a time, from your PC to your PLC, and test each circuit independently. The PLC can now function as an extremely useful test device.

If you are working on a new operation, obtain the PLC before it is installed and hard-wired into the control panel and set it up at your work area. If the new PLC is not available, obtain a PLC identical (or very similar) to it. This PLC can be reserved as a future spare, can be used for programming programmable read-only memory chips (PROMs), and can be used for circuit testing on this project and on future projects.

It is easy to use a PLC as a device for testing circuits. All you need to do is to provide power to the PLC and hard-wire several of the inputs. Three or four push buttons and a two-position or a three-position selector switch usually are sufficient. You do not have to hard-wire any of the outputs, because the output lights indicate whether an output is on or off.

When you use a spare PLC as a test device, it is no longer necessary to learn how to develop RLL circuits the old way. The old way involved development of a complete circuit for a machine or process. Copies of the circuit would be delivered to the maintenance department so that the electricians could hard-wire the machine. Only after the machine was completely hard-wired and testing started could the "bugs" be found. Then it was back to the drawing board.

Using a PLC at your work area and testing small segments of your circuit as you complete them, you can eliminate your bugs immediately. The trial-and-error system works very well in this situation, and your learning curve will improve very rapidly as you see what does and does not work.

As you work your way through the pages of this book, you will see example circuits. One of the main goals of this book is to provide many example circuits for you to copy into your programs or into an experimental circuit. I hope these examples reduce some of the discomforts of your trial-and-error efforts.

As this chapter closes, note that an exercise with a "Practice Circuit" is presented in chapter 4. This is a very complete exercise, and you will be taken through the complete cycle of constructing, testing, and revising a circuit. This exercise starts out in the section entitled "Practice Circuit," which includes the circuit to copy into the EX-PDD software in your PC. The second section is entitled "Validate." The third section is entitled "Going On-line" and covers the steps required to allow communications between the PC and the PLC. The fourth section is entitled "Automatic Identification of the I/O Allocation." The fifth section is entitled "Testing the Practice Circuit" and leads you through the steps required to "load" the circuit from your PC into your PLC. It also shows you how to operate the PLC and compare that operation with a written description of the functions of the circuit. And finally, the section entitled "Experimenting with the Practice Circuit" shows you how to make changes to the circuit. The exercise also leads you through the steps required to "save" any changes you made to the circuit and transfer those changes from the PLC back to the file in your PC.

2 TOSHIBA EX-PDD SOFTWARE

A FEW NOTES ABOUT EX-PDD VERSIONS

A few notes need to be made about the EX-PDD250 software versions that are in existence today.

The EX-PDD250 software currently being shipped from Toshiba is Version 3.X. The previous release was Version 2.X. Several changes have been incorporated into the version 3.X release. The most important changes are as follows:

1. Version 3.X includes mouse support.
2. Version 3.X includes an easy-to-use color setup for your monitor.
3. Version 3.X recommends that your personal computer (PC) be equipped with a 3.5-inch high-density floppy drive.

   Note: If necessary, there are ways to get around this requirement.
4. Version 3.X includes a documentation disk. This disk includes the manual for the EX-PDD250 software and the user manuals for the EX100 programmable logic controller (PLC), M20 PLC, and M40 PLC. This disk is formatted with Microsoft *Word for Windows* Version 2.0b. Also included on this disk is a .DXF file for computer-aided design (CAD) programs. It contains the wiring diagrams for the M20 PLC, the M40 PLC, and each EX100 input-output (I/O) module.
5. Version 3.X reads all Version 2.X programs. However, Version 2.X cannot read Version 3.X programs.
6. Version 3.X includes a Sentinel Key for the *serial* port of your PC. The Sentinel Key for Version 2.X was for the *parallel* port of your PC. The functions of the Sentinel Key protection have also been reversed.

Point 6 means that the Sentinel Key can be permanently fastened with the connector screws to the Converter Module cable. The Converter Module is the device that plugs into the serial port of your PC and the programming port of your PLC. This makes it easy to keep track of the Sentinel Key.

Point 6 also means that in Version 2.X (without using the Sentinel Key) you could do anything in the software that could be done off-line, but you could not go on-line to communicate with the PLC. In Version 3.X, you can now go on-line without using the Sentinel Key. You can even write or change a program on-line and then save the program in the PLC to the hard drive of your PC without using the Sentinel Key. However, you cannot write synonyms on-line; you cannot write programs, commentaries, or synonyms off-line; and you cannot print programs off-line without the Sentinel Key.

All of the releases of the EX-PDD250 software (with the exception of Versions 2.60 and 2.61) have been designed for PCs that have an eight-bit architecture. This means that the EX-PDD250 software can be run on any PC, starting with the XT style that contains an 8088 microprocessor, and continuing up through PC styles that contain the 80286, 80386, 80486, or Pentium microprocessors.

This book, then, is written primarily around Version 3.X of the EX-PDD250 software. However, an extremely high percentage of the information is relative to all of the earlier versions. In fact, I have included information about Version 2.X in the sections entitled "EX-PDD Software Quick Start—Version 2.X" and "EX-PDD Software Top Menu—Version 2.X."

A FEW NOTES ABOUT THE DEMONSTRATION DISK

The demonstration disk that accompanies the *Toshiba Medium PLC Primer* is a special version that combines features from Version 2.X and Version 3.X of the EX-PDD250 software. This demonstration disk is not a full-featured version like the Version 3.X that was described in the previous section. The demonstration disk has been designed especially for readers who do not yet have a PLC but who want to learn more about programming one before making the decision to purchase one. Therefore, the demonstration disk allows you to work with 95 percent of the *Toshiba Medium PLC Primer.*

The demonstration disk allows you to do the following:

1. Hand copy a relay ladder logic (RLL) program from the primer onto the screen of your PC.
2. Program an original RLL program of your own into your PC.
3. Save any RLL program from the screen to the hard drive of your PC.
4. Validate your program.

5. Add synonyms and descriptions to your devices.
6. Print a copy of your RLL program.
7. Use your mouse to construct your RLL program.
8. Use the color setup for your monitor.
9. Use the demonstration disk and Microsoft *Word for Windows* Version 2.0b to review the manual for the EX-PDD250 software as well as the user manuals for the EX100 PLC, the M20 PLC, and the M40 PLC.

What will you *not* be able to do with the demonstration disk that accompanies the *Toshiba Medium PLC Primer*? You will *not* be able to go on-line with either the Practice Circuit or any of your own programs. Let me repeat: the demonstration disk has been designed especially for readers who do not yet have a PLC but who want to learn more about programming one before making the decision to purchase one.

To go on-line with any programmed circuit, you must purchase a Toshiba Medium PLC, a Converter Module, and the full-featured EX-PDD250 Version 3.X software. The software comes with the Sentinel Key, which allows you to access the complete set of features.

Note: Throughout this book, characters and commands that you are to type are printed boldface.

EX-PDD SOFTWARE QUICK START—VERSION 3.X

It's time to install the EX-PDD software on the hard drive of your PC.

Note: Toshiba recommends that you not attempt to operate the software from a PC that is not equipped with a hard drive.

Note: Toshiba is shipping Version 3.X software on 3.5-inch high-density (HD) floppy disks only. If your PC is not equipped with a 3.5-inch high-density floppy drive, contact your distributor for a copy of the software on 5.25-inch HD floppy disks or on 3.5-inch double-sided, double-density (DSDD) floppy disks.

You must accomplish three tasks before installing the program. If you have never done any of these tasks, seek out your company's computer guru for assistance. If that is not possible, refer to your DOS manual. These tasks are as follows:

1. Make a backup copy of the EX-PDD disk. Store the original in a remote and safe place.
2. Modify the C:\CONFIG.SYS file to include the following:

FILES = 30
BUFFERS = 30

3. Create a new directory for the EX-PDD software. I recommend that you assign this directory a name such as PDD-250 or TOSH250. If you use one of these names, or a slight variation of it, you can later create a directory such as PDD-2040 or TOSH2040 if you work with the Toshiba series of small PLCs, which use a reduced form of EX-PDD. Because you are using DOS, you are limited to a directory name of eight or fewer characters.

Now load the software. There are now two methods of installing Version 3.X of the EX-PDD250 software into your PC. The first is with the Installation Utility, and the second is with DOS commands.

For this example, I have named the new directory PDD-250. To load the software, follow either of the next two methods.

Using the Installation Utility

1. At the C:> prompt, type **CD\PDD-250** and press ENTER.

   Note: This command changes the directory from the root directory to the new directory. The prompt changes from C:\> to C:\PDD-250>.
2. Using the newly made backup copy, put the disk into the A: drive.
3. At the C:\PDD-250> prompt, type **INSTALL A:*.***
4. Now follow the instructions on your screen.

   Note: This command copies the subdirectories and all of the files from the floppy disk to the hard drive and stores them in the new directory and subdirectories.
5. The EX-PDD is now loaded on your hard drive. To run EX-PDD, type **PDD** and press ENTER.

   You are on your way!

Using the DOS Commands

1. At the C:> prompt, type **CD\PDD-250** and press ENTER.

   Note: This changes the directory from the root directory to the new directory. The prompt changes from C:\> to C:\PDD-250>.
2. Using the newly made backup copy, put the disk into the A: drive.
3. At the C:\PDD-250> prompt, type **XCOPY A:*.* /S**

Note: This copies the subdirectories and all of the files from the floppy disk to your hard drive and stores them in the new directory and subdirectories.

4. The EX-PDD is now loaded on your hard drive. To run EX-PDD, type **PDD** and press ENTER.

You are on your way!

EX-PDD SOFTWARE QUICK START—VERSION 2.X

It's time to install the EX-PDD software into your PC. This section first describes installation for PCs equipped with a hard drive and then describes installation for PCs with floppy drives only.

Hard Drive

You must accomplish three tasks before installing the program. If you have never done any of these tasks, seek out your company's computer guru for assistance. If that is not possible, refer to your DOS manual. These tasks are as follows:

1. Make backup copies of all of the EX-PDD disks. Store the originals in a remote and safe place.
2. Modify the C:\CONFIG.SYS file to include the following:

   FILES = 30
   BUFFERS = 30
3. Create a new directory for the EX-PDD software. I recommend that you assign this directory a name such as PDD-250 or TOSH250. If you use one of these names, or a slight variation of it, you can later create a directory such as PDD-2040 or TOSH2040 if you work with the Toshiba series of small PLCs, which use a reduced form of EX-PDD. Because you are using DOS, you are limited to a directory name of eight or fewer characters.

Now load the software. For this example, I have named the new directory PDD-250. To load the software, follow these six steps:

1. At the C:> prompt, type **CD\PDD-250** and press ENTER.

   Note: This command changes the directory from the root directory to the new directory. The prompt changes from C:\> to C:\PDD-250>.

2. Using the newly made backup copies, put disk 1 into the A: drive.
3. At the C:\PDD-250> prompt, type **COPY A:*.* C:**

   Note: This command copies the files from disk 1 to your hard drive and stores them in the new directory.

4. When step 3 is complete, replace disk 1 with disk 2 and repeat step 3.
5. If your EX-PDD is on 5.25-inch disks, replace disk 2 with disk 3 and repeat step 3.
6. The EX-PDD is now loaded on your hard drive. To run EX-PDD, type **PDD** and press ENTER.

You are on your way!

Floppy Drive

You must accomplish two tasks before installing the program. If you have never done either of these tasks, seek out your company's computer guru for assistance. If that is not possible, refer to your DOS manual. These tasks are as follows:

1. Make backup copies of all of the EX-PDD disks. Store the originals in a remote and safe place.
2. Modify the CONFIG.SYS file to include the following:

   FILES = 30
   BUFFERS = 30

Using the newly made backup copies, put disk 1 into the A: drive, type **PDD**, and press ENTER.

You will receive the following message:

```
EX-PDD Install
Remove Disk 1 and insert Disk 2 into drive, then
Strike a key when ready . . .
```

Follow the instructions, and you are on your way!

EX-PDD SOFTWARE TOP MENU—VERSION 3.X

To use the EX-PDD software, start with the TOP MENU.

Once the EX-PDD software has been loaded into your PC or laptop computer, use the DOS commands to change to the directory that you have created. At the DOS prompt, type **PDD** and press ENTER. EX-PDD begins to load. If all is well, the next time your PC pauses, you will see a note near the bottom of the screen that reads as follows:

```
Press [ENTER] to start EX-PDD system.
```

Once you have pressed ENTER, the screen that appears is the TOP MENU. It looks like Figure 2.1. Figure 2.2 shows the Software Top Menu Tree for Version 3.X.

Figure 2.1 TOP MENU—Version 3.X

```
  Toshiba EX-Series Program Development and Documentation System (EX-PDD)
   for M20/M40, EX100, EX200B, EX250, and EX500 Programmable Controllers

[F1] - System and Control Information: [F6] - Offline/Online programming:
       PLC status and usage                   Toggle between offline (disk)
       Load/Save/Compare.                     ladder and online (PLC) ladder.

[F2] - Ladder Logic:                   [F7] - Enter/Change program name:
       Edit logic and commentary,             Specify ladder logic program;
       search, monitor and Force I/O.         program name is found on disk.

[F3] - Block Monitor:                  [F8] - Enter/Change comm. parameters:
       Display I/O device and                 Specify/Setup COM1 or COM2 for
       register values.                       PLC communication.

[F4] - Documentation:                  [F10]- Exit to DOS:
       Print ladder, commentary,              Temporary or permanent exit
       system information, and maps.          from EX-PDD program.

[F5] - User's Setting:                 Press [Ctrl F10], or click in the upper
       Color changes.                  left corner of a screen with the mouse
                                       at any time to access the HELP system.

Offline File name: default

 System  Ladder  Block   Documt  User's  Online  Program  COMM       1 Exit
1Infor. 2Logic  3Monitor4 Print 5Setting6Ladder 7 Name   8 Port 9    0 DOS
```

Figure 2.2 EX-PDD Software TOP MENU Tree—Version 3.X

```
[F1]    [F2]    [F3]     [F4]     [F5]     [F6]     [F7]     [F8] [F10]
System  Ladder  Block    Document User's   Online-  Program  COMM Exit
Info.   Logic   Monitor  Print    Setting  Offline  Name     Port to
Funct.  Funct.  Funct.   Funct.            Toggle                 DOS

|------ |------ |------- |------- |------- |------- |------- |--- |----
|       |       |        |        |        |        |        |     Temp.
|       |       |        |        |        |        |        |     Perm.
|       |       |        |        |        |        |        |
|       |       |        |        |        |        |        Select
|       |       |        |        |        |        |        Communicat
|       |       |        |        |        |        |        Parameters
|       |       |        |        |        |        |
|       |       |        |        |        |        Select Ladder
|       |       |        |        |        |        Filename
|       |       |        |        |        |
|       |       |        |        |        Online PLC Ladder Program.
|       |       |        |        |        Offline PC Ladder Program.
|       |       |        |        |
|       |       |        |        Color changes.
|       |       |        |
|       |       |        Define Printer Port
|       |       |        Select Output Options
|       |       |        Start Print
|       |       |
|       |       Enter Device and Timer/Counter Values
|       |       Display Device and Timer/Counter Values
|       |
|       PLC Programming
|       Power Flow Monitor
|       Ladder Commentary
|
PLC Status
Control PLC
Reset PLC
Load PLC from PC
Save PLC to PC
Compare File to PLC
Compare File to File
```

EX-PDD SOFTWARE TOP MENU—VERSION 2.X

To use the EX-PDD software, start with the TOP MENU.

Once the EX-PDD software has been loaded into your PC or laptop computer, use the DOS commands to change to the directory that you have created. At the DOS prompt, type **PDD** and press ENTER. EX-PDD begins to load. If all is well, the next time your PC pauses, you will see a note near the bottom of the screen that reads as follows:

```
Press [ENTER] to start EX-PDD system.
```

Once you have pressed ENTER, the screen that appears is the TOP MENU. It looks like Figure 2.3. Figure 2.4 shows the Software Top Menu Tree for Version 2.X.

Figure 2.3 TOP MENU—Version 2.X

```
Toshiba EX-Series Program Development and Documentation System
(EX-PDD) for M20/M40, EX100, EX200B, EX250, and EX500
Programmable Controllers

  F1 - System and Control Information:
       PLC status and usage, Load/Save/Compare.

  F2 - Ladder Logic:
       Edit logic and commentary, search, monitor and Force I/O.

  F3 - Block Monitor:
       I/O device and register values display.

  F4 - Documentation:
       Print ladder, commentary, system information, and maps.

  F6 - Offline/Online programming selection:
       Toggle between offline (disk) ladder & online (PLC) ladder

  F7 - Enter/Change program name:
       Specify the ladder logic program name to be found on disk.

  F8 - Enter/Change Communication parameters:
       Specify/Setup COM1 or COM2 for PLC communication.

 F10 - Exit to DOS:    (Insert DOS disk.)
       Temporary or permanent exit from EX-PDD program.

[Ctrl] F10 - HELP information. (Insert HELP disk.)


Offline File name: default

 System  Ladder  Block   Documt   Online  Program  COMM   1 Exit
1Infor. 2Logic  3Monitor4 Print 5 6Ladder 7 Name  8 Port 9 0 DOS
```

Figure 2.4 EX-PDD Software TOP MENU Tree—Version 2.X

```
[F1]     [F2]     [F3]     [F4]      [F6]     [F7]     [F8]  [F10]
System   Ladder   Block    Document  Online-  Program  COMM  Exit
Inform.  Logic    Monitor  Print     Offline  Name     Port  to
Funct.   Funct.   Funct.   Funct.    Toggle                  DOS

|------  |------  |------  |------   |------  |------  |---  |----
|        |        |        |         |        |        |     Temp.
|        |        |        |         |        |        |     Perm.
|        |        |        |         |        |        |
|        |        |        |         |        |        Select
|        |        |        |         |        |        Communicat.
|        |        |        |         |        |        Parameters
|        |        |        |         |        |
|        |        |        |         |        Select Lad. Filename
|        |        |        |         |
|        |        |        |         Online PLC Ladder Programming
|        |        |        |         Offline PC Ladder Programming
|        |        |        |
|        |        |        Define Printer Port
|        |        |        Select Output Options
|        |        |        Start Print
|        |        |
|        |        Enter Device and Timer/Counter Values
|        |        Display Device and Timer/Counter Values
|        |
|        PLC Programming
|        Power Flow Monitor
|        Ladder Commentary
|
PLC Status
Control PLC
Reset PLC
Load PLC from PC
Save PLC to PC
Compare File to PLC
Compare File to File
```

EX-PDD SOFTWARE COMMAND LINE MENUS

This section describes several frequently used command lines that are part of the EX-PDD software. These command lines appear at the bottom of your screen and are accessed by pressing one of the Function (F) keys that are located across the top of, or down the left side of, your keyboard.

Note: The command lines shown in this chapter are not exactly as they appear on your screen. They have been compressed to fit the printed page.

Figure 2.5 Command Line 1

```
Offline File name: default

Display New  Global  Edit Delete  Search  Valid  Comment   1 Top
1 Page 2Page 3 Page 4Page 5 Page 6Ladder 7Ladder 8Ladder 9 0 Menu
```

Command Line 1 All commands in command line 1 (Figure 2.5) do what they say they do. For example, if you press F4, you are put in the Edit Page or "Create a Circuit" mode of programming. Activating any of these commands brings up a new set of command lines. The only exception to this is F7–Valid Ladder. If you press F7, the software validates the page at which you are looking. If you press Shift-F7 (press and hold down the Shift key while pressing F7), the software validates the complete program automatically and sequentially. By the way, the Shift-F7 command is not described in the EX-PDD manual. Validating the ladder is discussed in chapter 4 in the section entitled "Validate."

Command Line 2 Almost all the commands in command line 2 (Figure 2.6) do what they say they do. However, in many of the commands, the top half of the line may mean something altogether different than the bottom half of the line. Or the command on the top half of the line may be a variation of the command on the bottom half of the line. For example, F1 through F8 are as they seem. Pressing F1–Seq. Ins brings up command line 3, which contains a number of sequence instructions (see next section). F2 through F6 bring up various advanced instructions, such as F3–Move and Compare or F5–Logic Operations.

F9–Assign Synonym has two separate functions. Pressing F9 changes the command from Assign Synonym to Assign Device. This command allows you to toggle between assigning a synonym and assigning an address to a device. However, with Shift-F9, a three-line chart is inserted at

Figure 2.6 Command Line 2

```
Offline File name: default

 Seq. Ext.  Move        Logic  Conv.  Insert  Move  Assign  1Acc.
1Ins 2Seq. 3Comp 4Arith 5Oper 6Limit 7Delete 8Copy 9Synonym 0
```

Figure 2.7 Command Line 3

```
Offline File name: default

 -]^[-  -]v[-  OPEN   -x x-  FORCE  TOF  SS        Assign 1Accept
1-] [- 2-]/[- 3CLOSE 4----- 5-( )- 6TON 7CNT 8END 9Device 0 Main
```

the bottom of the screen so that you can assign addresses, synonyms, and descriptions to all of your devices.

F10–Accept means the same as SAVE, and you must save every page before leaving it. However, ACCEPT is on the top command line, and therefore you must use Shift-F10 to accept the page of the program you have just completed.

Command Line 3 The following comments apply to command line 3 (Figure 2.7).

F1	-] [-	Normally open (NO) input contact.
Shift-F1	-]^[-	Rising edge transitional contact. Use only sparingly (see chapter 8, tip 5).
F2	-]/[-	Normally closed (NC) input contact.
Shift-F2	-]v[-	Falling edge transitional contact. Use only sparingly (see chapter 8, tip 5).
F3	CLOSE	Produces a short vertical line for use in creating parallel branches of a circuit.
Shift-F3	OPEN	Erases a short vertical line that was created by F3–CLOSE.
F4	—	Deletes an unwanted device and replaces it with a solid line.
Shift-F4	-x x-	Creates a gap in the ladder rung.
F5	-( )-	The coil for an output.
Shift-F5	FORCE	Allows you to energize a coil and keep it energized for testing or debugging purposes.
F6	TON	ON-Delay timer function.
Shift-F6	TOF	OFF-Delay timer function.
F7	CNT	Two-rung counter function.
Shift-F7	SS	Single-shot timer.
F8	END	END statement.
F9	DEVICE	Allows you to toggle the cursor (while in the ladder diagram) to a position below a device to insert an address or to a position above the device to insert a synonym.
Shift-F9	ASSIGN	Inserts a three-line chart at the bottom of the screen so that you can assign addresses, synonyms, and descriptions.

F10	MAIN	Returns you to command line 2.
Shift-F10	ACCEPT	ACCEPT or SAVE command.

SYNONYMS AND DESCRIPTIONS

This section describes the procedure for entering *device commentary synonym* and *device commentary description* notes into your program. Using the instructions in the section entitled "EX-PDD Software Quick Start," start EX-PDD and proceed as follows:

1. From the TOP MENU, select F7–Enter/Change program name.
2. While at the Enter/Change program name screen, type over the word *default* with the word **PRACTICE** or with the program name of your choice.
3. Press F10 to save that name and bring you back to the TOP MENU.
4. From the TOP MENU, go to command line 1 by pressing F2–Ladder Logic.
5. At command line 1, press F8–Comment Ladder.
6. Step 5 presents you with another command line; press F8–Device Comment.

The top of the screen looks like Figure 2.8.

The cursor is automatically placed under REF, so type **R000**. Then use the TAB key to move the cursor under SYN, and type **START**. Tab past the COLOR column to the TOP column, and type **START**. Tab over to the MIDDLE column, and type **INTERN**. Tab over to the BOTTOM column, and type **RELAY**. Press ENTER to save that line of type.

Your screen should look like Figure 2.9.

Remember to press ENTER after you have typed all of the information on each line. After you have pressed ENTER to save a line of addresses, synonyms, and descriptions, the cursor is automatically placed on top of

Figure 2.8 Initial Address, Synonym, and Description Screen

```
      REF          SYN       Color       Top       Middle     Bottom

                              000
                              000
=>                            000                                  <=
                              000
                              000
```

Figure 2.9 A Line of Addresses, Synonyms, and Descriptions for the "Practice Circuit"

REF	SYN	Color	Top	Middle	Bottom	
		000				
		000				
=> R000	START	000	START	INTERN	RELAY	<=
		000				
		000				

the first digit in the line that you have just completed. Leave the cursor there and type **R001**, or whatever your next address happens to be.

It will appear that you are erasing the previous address; but that is not the case. When you press the TAB key again, a new line is established with address R001 as the REF and empty spaces for your new synonym and description. The line that you just finished has simply been pushed up one line. The program shows only five lines of information on the screen at one time. The other lines are scrolled off at the top or the bottom but can be accessed with up or down cursor movements.

This program keeps track of your addresses in alphanumeric order. In other words, *R*'s (relays) come before *X*'s (inputs), and *X*'s come before *Y*'s (outputs). Therefore, it does not matter where you are in the list when you try to type in a new address, because the program keeps your list in order.

Now, type six more rows as shown in Figure 2.10.

Figure 2.10 Additional Addresses, Synonyms, and Descriptions for the "Practice Circuit"

REF	SYN	Color	Top	Middle	Bottom	
R001	RUN	000	RUN	INTERN	RELAY	
X001	PBSTRT	000	START	PUSHBT	(mom.)	
Y020	ALARM	000	ALARM	LIGHT		
Y021	LIGHT1	000	LIGHT	NO. 1		
Y022	LIGHT2	000	LIGHT	NO. 2		
=> Y023	LIGHT3	000	LIGHT	NO. 3		<=

To exit this screen, press F10 two times.

A convenient feature results from the exercise that you just completed. While you are programming a circuit, and after you have entered a device, typed in the address below the device, and pressed the ENTER key to save the address, the synonym is immediately displayed above the device.

If you prefer, after you have entered a device, you can press the F9–Assign Device key so that the cursor flips to the top of the device. Then you can type in the synonym and press the ENTER key to save the synonym, and the address is immediately displayed below the device.

When you use longer programs than are described in this book and you assign numerous synonyms and descriptions, you need to TEST and REPACK the indexes. These two features also appear on the synonym and description screen. When you have finished a series of synonyms and descriptions, press F2–Test Index. A message tells you if the test has passed or failed. If you receive the following message:

```
REF (or SYN) Index Failed-Time to Rebuild Indexes
```

Then press F1–Repack Index and wait. Do this for both the REF and SYN categories.

Do not wait until you have entered all of your lines of synonyms and descriptions before performing the TEST and REPACK operation. If you find the program slowing down as you type, don't wait any longer; perform the TEST and REPACK operations immediately.

Note that the EX-PDD software does not accept the use of the same synonym twice. If you try to enter a synonym a second time, you will receive a message that the synonym is invalid. This usually means you have used that synonym for another device.

3 BASIC RLL OPERATIONS

ADDRESSES

The concept of addresses is extremely important, yet it is a concept that is often the hardest for first-time users of programmable logic controllers (PLCs) to grasp.

The address is the device that ties together all elements of the PLC. It ties an input or an output to its own terminal for hard-wiring; it ties the software command to that particular terminal; and it ties the synonyms and descriptions to that particular input or output.

Picture your PLC as if it were your checking account. Your checking account has a number (or address). When you make a deposit, all of the funds go to the checking account with that particular number or address. When a check gets to the bank, the money is withdrawn from the account with that particular address. If you should attempt to deposit some cash into *your* account, but use the deposit slip from your *spouse's* account—well, you get the idea. That's how important addresses are.

Besides having a synonym and a description, every device you use in your program must have an address. This applies to the inputs and the outputs that are external and physical devices, as well as to the relays, timers, and counters that are the internal, software-created devices.

The address for an input or an output is the same as the actual terminal to which you wire the device. These addresses are labeled on the PLC. Addresses for the relays, timers, and counters are especially helpful in preventing duplicate uses of the devices, which can wreak havoc in your program.

The most common addressing systems are binary, octal, decimal, and hexadecimal. *Binary* means two, *octal* means eight, *decimal* means ten, and *hexadecimal* means sixteen. Decimal is, of course, the system with which we all are familiar—the standard one through ten. And we have all come to recognize the binary zero and one through our use of computers. Toshiba uses the octal system in the Small PLC series and the hexadecimal system in the Medium PLC series.

Let me mention, in passing, that the octal system uses the following sequence: 0, 1, 2, 3, 4, 5, 6, and 7. But we need to explore the hexadecimal system in some depth. First of all, the numbering system starts with 0. All devices can have an address of 0. It can be X000 for an input; it can be Y000 for an output; it can be R000 for an internal relay; it can be T000 for a timer; and it can be C00 for a counter. The hexadecimal system uses the following sixteen digits: 0, 1, 2, 3, 4, 5, 6, 7, 8, 9, A, B, C, D, E, and F. Don't let this numbering system confuse you. However, it does take practice to get used to it.

To get started, the following is a four-level hexadecimal address system:

00, 01, 02, 03, 04, 05, 06, 07, 08, 09, 0A, 0B, 0C, 0D, 0E, 0F.

10, 11, 12, 13, 14, 15, 16, 17, 18, 19, 1A, 1B, 1C, 1D, 1E, 1F.

20, 21, 22, 23, 24, 25, 26, 27, 28, 29, 2A, 2B, 2C, 2D, 2E, 2F.

30, 31, 32, 33, 34, 35, 36, 37, 38, 39, 3A, 3B, 3C, 3D, 3E, 3F.

You will find yourself skipping the A, B, C, D, E, and F digits. This does not present a great problem when you are dealing with the addresses for the internal relays. However, if you skip these digits when dealing with addresses for inputs and outputs, you will find that when you try to hard-wire your PLC, 37.5 percent of your input and output terminals will not have devices assigned to them. Therefore, it is definitely easiest to remember the hexadecimal numbering system when dealing with the inputs and the outputs of your PLC, because the input and the output terminals are labeled that way.

INPUTS

The two main types of devices that are associated with PLCs are inputs and outputs. These devices are external to the PLC and are found somewhere on your equipment.

Consider inputs as Change of State devices. This means that regardless of the way in which an input device is wired or programmed, the device always reverses its condition whenever the input device is operated.

Inputs and outputs are always hard-wired to the PLC at the appropriate terminals. *A word of caution*: inputs are wired differently from outputs, and the various models available in the Toshiba line of PLCs have different wiring schemes for both inputs and outputs. Be careful wiring the PLC. Follow the wiring diagram that applies to your particular model for that input or output type.

Inputs come in three varieties: dry contact input, alternating current (AC) input, and direct current (DC) input. The AC and the DC inputs also come in a variety of voltages.

The *dry contact* configuration does not require any external source of power to operate the input devices; the power is supplied by the PLC itself, in the form of 24 volts of direct current (VDC). However, both the *AC input* and *DC input* require that you provide a power source. One side of the source is connected to the C terminal; the other side is used as a supply voltage for each input device.

The inputs you will use come in a wide variety of devices. The following are some of the types:

Float switches	Motion sensors
Flow switches	Photo-eye sensors
Level switches	Pressure switches
Limit switches	Selector switches
Maintained push buttons	Temperature switches
Momentary push buttons	Thumbwheel switches

It does not matter what type of device you are using as the input, as long as you assign it the same address as the PLC terminal to which you wire it. It also helps if you assign a synonym and description to the input to fully describe it.

For example, you have three limit switches in your system, and you assign the input addresses X000, X001, and X002. So far, so good. When you start to hard-wire these inputs to your PLC, it is extremely helpful if you have assigned synonyms and descriptions. For example, the following assignments could be made:

Address	*Synonym*	*Description*
X000	CON LS	CONVEY LIMIT SWITCH
X001	ELE LS	ELEVAT LIMIT SWITCH
X002	JAM LS	JAM LIMIT SWITCH

A copy of this list can be printed with the documentation portion of the software and be given to the person who wires the PLC.

In the past, circuit designers have had to represent each input with a different symbol. A symbol that looked like a float represented a float switch. A push button was represented by a symbol that looked like a pair of contacts that was either next to, or away from, another set of contacts.

Input symbols for PLCs are limited to two types—normally open (NO) and normally closed (NC). The symbol for the NO contact is -| |-, and the symbol for the NC contact is -|/|-.

Unless there is an excellent reason to do otherwise, all inputs should be hard-wired NO, regardless of what the relay ladder logic (RLL) program shows. If you follow this rule, there should never be a misunderstanding between the person doing the programming and the person doing the hard-wiring. This rule allows the circuit designer or programmer to set the input to either NO or NC as required by the system.

A final note about inputs. When designing your control system, also provide an additional fuse or circuit breaker that will handle all of the inputs.

OUTPUTS

Once again, the two main types of devices that are associated with PLCs are inputs and outputs. These devices are external to the PLC and are found somewhere on your equipment.

Inputs and outputs are always hard-wired to the PLC at the appropriate terminals. *A word of caution*: outputs are wired differently from inputs, and the various models available in the Toshiba line of PLCs have different wiring schemes for both inputs and outputs. Be careful wiring the PLC. Follow the wiring diagram that applies to your particular model for that input or output type.

There are three different types of outputs—relay, triac, and transistor. All three output configurations require some type of power.

Note: Some of the symbols in the Toshiba manuals are not clear on this point, especially where 120 VAC is required. Sometimes the symbol used is a circle with the letters *PS* inside it. Sometimes the circle contains the tilde (~) symbol, which signifies an AC power source. Sometimes, the circle is empty, and that is where the problem arises. Just remember that all outputs require a power source.

The outputs you will use come in a wide variety of devices. The following are some of the types:

AC variable frequency drives	Pilot lights
Alarms (all types)	Pumps
Fans	Relays (large)
Heaters	Solenoids
Motors (small)	Solid state starters
Motor starters	Valves

The symbol for the PLC output is -( )-.

It does not matter what type of device you are using as the output, as long as you assign it the same address as the PLC terminal to which you wire it. It also helps if you assign a synonym and description to the output to fully describe it.

For example, you have three outputs in your system, and you assign the output addresses Y020, Y021, and Y022. So far, so good. When you start to hard-wire these outputs to your PLC, it is extremely helpful if you have assigned synonyms and descriptions. For example, the following assignments could be made:

Address	*Synonym*	*Description*
Y020	CON MS	CONVEY MOTOR STARTR
Y021	JAM PL	JAM PILOT LIGHT
Y022	FAN	FAN

A copy of the list can be printed with the documentation portion of the software and be given to the person who wires the PLC.

Make sure that you pay particular attention to the amperage that each output device draws. Check the specifications of the device, and then check the amperage rating of your PLC (see "Specifications" in your PLC manual). Do not create a current overload on the PLC output. If your load device draws more current than the maximum the PLC output allows, install an external (or interposing) relay in your system that can be safely handled by the PLC output and that can handle the current draw of your load device.

A final note about outputs. When designing your control system, and also when sizing your control panel, provide a fuse or circuit breaker for each of the outputs. Providing protection for each output at the start of the project can save many hours of troubleshooting in the future, especially when you try to find the source of a blown fuse if eight, ten, or even more outputs are on a single fuse. One excellent method of providing this protection is to install terminal strips that have built-in fuse holders.

THE END STATEMENT

Every circuit needs an END statement—the program will not run without one. When you believe you have completed each program, install an END statement. This END statement can be the last item on the last page of your circuit. However, it is highly recommended that you create one extra page and that the END statement be the only item on that page.

Figure 3.1 Last Page of the Circuit

```
| +-----+
|-| END |
| +-----+
```

There is an excellent reason for this recommendation. Sometime in the future, you may decide to revise and expand the original circuit that you developed. If your END statement is the last device on the last page of the original circuit, you could easily forget to remove that original END statement. If you add the new portions of your circuit and another END statement, but fail to remove the original END statement, problems will arise. If you have two END statements in your circuit, the PLC executes only the circuit that comes before the first END statement. It ignores the portion of the circuit between the first and the second END statements.

The END statement is critical to the scan of the program. The scan begins at the start of the circuit and looks at each device on each page until it comes to the END statement. It is at this time that the PLC makes the necessary changes to the outputs, based on the changes encountered during the scan that are different from those encountered in the previous scan.

Enter the END statement by pressing F8–END from command line 3. The final page of your circuit should look like Figure 3.1.

An END statement is *not* included in all of the circuits in this book. If you decide to copy any circuits for practice, testing, or inclusion in one of your circuits, remember to add the END statement.

TIMERS

The Toshiba Medium PLCs can access on-delay timers, off-delay timers, and single-shot timers. The software accepts time values of five digits, from 00001 to 32767. But be forewarned. The time values are in tenths of a second (0.1) or in hundredths of a second (0.01) depending on the timer address you select. Therefore, a value that you enter as 12000 could mean 1200.0 seconds, or it could mean 120.00 seconds. But it will *never* mean 12,000 seconds.

Let me take some of the confusion out of these timer values. A total of 128 timers are available in the medium series of PLCs. Eight of these tim-

Figure 3.2 An On-delay Timer Alone on a Partial Rung

```
|
|------------------------[          TON          ]
|
```

Figure 3.3 An On-delay Timer on a Partial Rung with Other Inputs

```
|RUN     PROCES
|-| |----| |---[          TON          ]
|R002    R00A
```

ers have a more precise time control. The addresses for these timers, along with their time values, are as follows:

Timer Addresses	*Time Values*
T000 through T119	0.1 seconds
T120 through T127	.01 seconds

The addresses for the timers are in the decimal numbering system, not in the hexadecimal system.

You can access the three different types of timers (on-delay, off-delay, and single-shot) from command line 3 by pressing F6–TON, Shift-F6–TOF, or Shift-F7–SS.

The timers for the medium series of PLCs are different from those of the small series. In the Medium PLCs, the timers are complete devices. When a timer is called up, the software puts it wherever you place the cursor on the rung. It looks like Figure 3.2.

If you have other inputs on the rung, it looks like Figure 3.3.

The cursor will also appear, and is positioned in the left-hand portion of the bracket. This is where you type in the time value you want to specify (Figure 3.4).

After typing in the time value, press ENTER. The software inserts the extra zeros, and your entry is changed to look like Figure 3.5.

Figure 3.4 Location of Time Value for an On-delay Timer

```
|RUN     PROCES
|-| |----| |---[300      TON          ]
|R002    R00A
```

Figure 3.5 Location of Time Value for an On-delay Timer after Keystrokes Have Been Entered

```
|RUN     PROCES
|-| |----| |---[00300   TON          ]
|R002    R00A
```

Figure 3.6 Location of Address for an On-delay Timer after Keystrokes Have Been Entered

```
|RUN     PROCES
|-| |----| |---[00300   TON    T013 ]
|R002    R00A
```

Figure 3.7 Timer Circuit after Coil Has Been Added

```
|RUN     PROCES                                                  BLOWER
|-| |----| |---[00300   TON    T013 ]---------------------------( )-
|R002    R00A                                                    Y026
```

Move the cursor two spaces to the right, to the right-hand portion of the bracket. Type your timer address here, and press ENTER again. The configuration shown in Figure 3.6 appears.

The timer value and address is complete, but you still need to finish the circuit with a coil. The coil can be an output device, or it can be an internal relay. It may look like Figure 3.7.

This circuit turns on the BLOWER (Y026) 30.0 seconds after the RUN (R002) and the PROCES (R00A) coils (elsewhere in the program) have been energized.

COUNTERS

All Toshiba Medium PLCs provide counters and up/down counters. This book discusses only the counters. The up/down counter is listed as Fun. No. 111 in the Other Functions section of the chapter entitled "Instructions" in your PLC manual. You are welcome to explore the world of up/down counters at your leisure. The software accepts count values of five digits, from 00001 to 65535.

You can access the counter from command line 3 by pressing F7–CNT. The counters for the medium series of PLCs are different from those of the small series. In the Medium PLCs, the counters are complete de-

Figure 3.8 A Counter

```
+-----------+
C  CNT      Q
|           |
|           |
|           |
E           |
+-----------+
```

vices. A counter requires two rungs. When a counter is called up, the software puts it wherever you place the cursor on the rung. It looks like Figure 3.8.

Because both the COUNT (C) input and the RESET or ENABLE (E) input require contact inputs to operate, you will frequently encounter the arrangement in Figure 3.9.

Two operations must be performed on this counter. First, a preset count value must be assigned. Move the cursor to the space just to the right of the E, and type in a value, **5** for example. Press ENTER to save this value, and the value is changed to 00005. Move the cursor one position to the right, and type in **C00**. Press ENTER again to save the address. This completes the counter. It looks like Figure 3.10 when you have finished.

The addresses for the ninety-six counters are in the decimal numbering system (C00 to C95), not in the hexadecimal system.

The C input accepts a pulsed input (it can be a fast or a slow pulse). However, to accept any pulse, the E input also must have a signal. When the E input is off, the C input pulses will *not* be accepted as counts. When the number of pulses counted equals the pre-set count value, the counter output Q is energized.

The counter program is complete, but you still need to complete the Q output circuit of the counter with a coil. It can be an output device, or it can be an internal relay. The completed counter circuit looks like Figure 3.11.

Figure 3.9 Counter with Count (C) and Enable (E) Inputs

```
|          +-----------+
|-|  |---C  CNT      Q
|          |           |
|          |           |
|          |           |
|-|  |---E           |
|          +-----------+
```

Figure 3.10 Completed Counter

```
|          +-----------+
|-| |---C   CNT        Q
|          |           |
|          |           |
|          |           |
|-| |---E 00005 C00    |
|          +-----------+
```

Figure 3.11 Completed Counter Circuit

```
|PHEYE   +-----------+                                          ALARM
|-| |---C   CNT      Q -----------------------------------------( )-
|X01B    |           |                                          Y027
|        |           |
|START   |           |
|-| |---E 00005 C00  |
|R000    +-----------+
```

This complete counter circuit might be used to count parts that have been rejected off of a main conveyor line. The rejected parts are sent to a gravity conveyor to be accumulated for reworking. As the parts move past photo-eye X01B, each rejected part sends a pulse to the counter. R000 is part of the main conveyor system, so the counter is on as long as the main conveyor is running. When five rejected parts have been counted, the counter activates ALARM Y027. The signal alerts the operator that a problem may exist because of the quantity of parts that have been rejected.

Remember that in order to accept any pulsed signal, the E input also must have a signal. Whenever the E input is off, the input pulses are *not* accepted as counts.

FLIP/FLOPS

A *flip/flop* is an internal PLC device that, once set, stays in that condition until it is reset.

The following is a truth table for a flip/flop:

Set Input	*Reset Input*	*Output Status*
ON	OFF	ON
OFF	ON	OFF
ON	ON	OFF
OFF	OFF	UNCHANGED

To insert a flip/flop into your diagram, go to command line 2, press F2–Extra Sequence, and select F6–F/F from the next command line. The flip/flop command inserts both the SET INPUT (S) and RESET INPUT (R) rungs to form the complete device at one time. You must modify the two input lines (S and R) to insert the contacts that will make the device work correctly. Depending on the type of circuit with which you are working, you may or may not elect to complete the Q output circuit of the flip/flop with a coil. The Q output can be an output device, or it can be an internal relay. The circuits in this chapter show both types of applications.

What follows are three different variations of flip/flop circuits. These three circuits use a three-position selector switch (SS 1A, SS 1B, and SS 1C) to test the three circuits independently.

Flip/flop Output 1 "Flip/flop Output 1" (Figure 3.12) can be found in all of the Toshiba PLC manuals. It requires two push-button inputs, and cycles the single output (OUT 0) on and off with alternate operations of the START and STOP push buttons.

Flip/flop Output 2 "Flip/flop Output 2" (Figure 3.13) represents a variation of the circuit presented in "Flip/flop Output 1." You can test it by rotating the selector switch to the second, or B, position. At that time, OUT 2 (Y022) comes on. It remains on until the START push button is pressed, at which time OUT 2 (Y022) goes off and OUT 1 (Y021) comes on. Repeated operation of the lone START push button causes the outputs to alternate on and off.

I usually recommend that the use of transitional contacts be avoided. However, this circuit does require one. In fact, it does not work without it. Transitional contacts and flip/flops use addresses that begin with an R, just like the standard internal PLC relays. To avoid confusion, it is a good practice to use addresses for these two special devices that are well above the range of addresses you normally use. Use addresses like R101 and

Figure 3.12 Flip/flop Output 1

```
SS 1A   START   +----+                                    OUT 0
-| |--+-| |---S FF Q-------------------------------------( )-
X005  |X000     |    |                                    Y020
      |         |    |
      |STOP     |    |
      +-| |---R R110
        X001    +----+
```

Figure 3.13 Flip/flop Output 2

```
|SS 1B   START                  +----+
|-| |--+-| |----|^|--+-|/|---S FF Q
|X006   |X000    R101   |R111     |   |
|       |               |         |   |
|       |               |         |   |
|       |               +-| |---R R111
|       |                 R111    +----+
|       |
|       |                                              OUT 1
|       +-| |----------------------------------------( )-
|       |R111                                          Y021
|       |
|       |                                              OUT 2
|       +-|/|----------------------------------------( )-
|        R111                                          Y022
```

R111, which are used in this circuit. Each transitional contact requires a specific address, such as R101. Also, the transitional contact with that address can be used only once in the program.

A typical use of "Flip/flop Output 2" might be where a limit switch is used to detect shipping cases and divert every other one from the main conveyor to alternate discharge chutes. Or it might be used with a pressure sensor or a level sensor to operate alternating pumps. Or it could be used to indicate whether a press is in the up or the down position.

Flip/flop Output 3 "Flip/flop Output 3" (Figure 3.14) is a modification of "Flip/flop Output 2." The X000 START contact has been deleted and has been replaced by internal relay R624, which is a special relay in the PLC that has a continuously operating time delay of 1.0 second (see

Figure 3.14 Flip/flop Output 3

```
|SS 1C                          +----+
|-| |--+-| |----|^|--+-|/|---S FF Q
|X007   |R624    R102   |R112     |   |
|       |               |         |   |
|       |               |         |   |
|       |               +-| |---R R112
|       |                 R112    +----+
|       |
|       |                                              OUT 3
|       +-| |----------------------------------------( )-
|       |R112                                          Y023
|       |
|       |                                              OUT 4
|       +-|/|----------------------------------------( )-
|        R112                                          Y024
|
| +----+
|-| END|
| +----+
```

the section in chapter 5 entitled "Flashing Light Circuits" for a complete description of this feature). When you rotate the selector switch to the third, or C, position, the alternating of OUT 3 (Y023) and OUT 4 (Y024) begins immediately, and they alternate at the 1.0-second rate.

SHIFT REGISTERS

All Toshiba Medium PLCs provide shift registers for your circuits.

Note: If you plan to test or experiment with the circuit presented in this section, you need three input devices—a two-position selector switch and two push buttons. Connect one of the NO contacts of the selector switch to input X007, connect the NO contact of one push button to input X000, and connect the NO contact of the second push button to input X001.

Shift registers are set up essentially the same way as counters and flip/flops. However, shift registers have three rung sections, and all three rungs appear on the screen at one time. To insert a shift register into your diagram, go to command line 2, press F2–Extra Sequence, and select F8–SR from the next command line. The shift register command inserts the DATA INPUT (D), the SHIFT INPUT (S), and RESET or ENABLE INPUT (E) rungs to form the complete device at one time.

The basic shift register appears on your screen wherever you place the cursor, and it looks like Figure 3.15.

Figure 3.15 Basic Shift Register

```
|              +----+
|--------------D SR Q
|              |    |
|              |    |
|              |    |
|--------------S(  )|
|              |    |
|              |    |
|              |    |
|--------------E    |
|              +----+
```

What follows is an explanation of the D, S, and E inputs. Don't let the descriptions overwhelm you. An example circuit follows these descriptions and it will allow you to test and experiment with the shift register. The example circuit will also help you to understand the meaning of these descriptions and show you just how a shift register works.

To begin with, the *D* stands for *data*. The only types of data that can be input to a shift register are ones and zeros. This means that the data input device is a contact, and it is either an external input device or an internal relay. Therefore, a closed contact presents an input signal of one (1) to the data input rung, whereas an open contact presents an input signal of zero (0) to the data input rung.

The *S* stands for *shift input.* The shift input must be on for the data to be shifted.

The *E* stands for *enable input. Enable* means the same as *reset.* The ENABLE input must be on for the shift register to operate. When the ENABLE input is off, all devices within the shift register are reset to off.

Let's talk about shift register addresses. First, the addresses are in the hexadecimal numbering system. Second, a shift register may contain up to sixty-four addresses. However, the address the shift register receives as its identifying address is that of the first address in the series that you are using.

The example circuit in this section uses a shift register with eight addresses. The identifying address is R016, which means that the series contains the following addresses: R016, R017, R018, R019, R01A, R01B, R01C, and R01D.

You need to realize that the sixty-four addresses that may be used in a shift register can control sixty-four separate devices. In the circuit in this chapter, R016 through R01D are the addresses for the contacts operated by the shift register.

Let's go back to our basic shift register and insert the information about the addresses. Place the cursor to the right of the S, and type (within the brackets) the number of addresses included in the series. In this circuit, that number is **8**. Press ENTER. Move the cursor down to the right of the E, type **R016**, and press ENTER again. Your shift register should now look like Figure 3.16.

In the second paragraph of this section, we mentioned a selector switch and two push buttons. Now insert an NC contact, for the selector

Figure 3.16 Shift Register after Addresses Have Been Inserted

```
|                  +----+
|------------------D SR Q
|                  |    |
|                  |    |
|                  |    |
|------------------S(08)|
|                  |    |
|                  |    |
|                  |    |
|------------------E R016
|                  +----+
```

switch with the address X007, in the D rung. Insert a NO contact, for the first push button with the address X000, into the S rung. Insert a NC contact, for the second push button with the address X001, into the E rung.

Make a note about the SHIFT input. An input of data from the DATA input occurs at each and every scan if the SHIFT input is closed. Because the scan time of your circuit is probably only about 2 or 3 milliseconds, you will get a great deal of data that you do not want. Therefore, a transitional contact is required in the SHIFT rung, which allows only the signal from the first scan to be accepted.

Your shift register circuit should now look like Figure 3.17.

The next step is to build a circuit for the eight contacts (R016 through R01D) for this shift register so that you can see how the shift register operates. Each of the eight addresses between R016 and R01D is used, so each contact is used to operate a separate output. This expands the circuit to look like Figure 3.18.

For the first demonstration, we will use only the X000 SHIFT push button to show how this shift register circuit works. Each time you press the X000 SHIFT push button, one more output light comes on. When you have energized all of the output lights, you can continue to press the SHIFT push button, and nothing happens. Or so it appears.

But let's go back to the beginning.

First, reset the circuit back to the original condition by pressing the X001 RESET push button. This clears the shift register and shuts off all of the output lights.

Press the X000 SHIFT push button once. The Y020 output comes on, because contact R016 has been energized, and it energized Y020. So far, so good.

Figure 3.17 Shift Register after Addition of Selector Switch and Two Push Buttons

```
|DATA              +----+
|-|/|----------D SR Q
|X007              |    |
|                  |    |
|SHIFT             |    |
|-| |----|^|---S(08)|
|X000    R253      |    |
|                  |    |
|RESET             |    |
|-|/|----------E R016
|X001              +----+
```

Figure 3.18 Completed Shift Register Circuit

```
|DATA           +----+
|-|/|----------D SR Q
|X007           |    |
|               |    |
|SHIFT          |    |
|-| |----|^|---S(08) |
|X000    R253   |    |
|               |    |
|RESET          |    |
|-|/|----------E R016
|X001           +----+
|
|-| |---------------------------------------------------( )-
|R016                                                  Y020
|
|-| |---------------------------------------------------( )-
|R017                                                  Y021
|
|-| |---------------------------------------------------( )-
|R018                                                  Y022
|
|-| |---------------------------------------------------( )-
|R019                                                  Y023
|
|-| |---------------------------------------------------( )-
|R01A                                                  Y024
|
|-| |---------------------------------------------------( )-
|R01B                                                  Y025
|
|-| |---------------------------------------------------( )-
|R01C                                                  Y026
|
|-| |---------------------------------------------------( )-
|R01D                                                  Y027
|
| +----+
|-| END|
| +----+
```

Press the X000 SHIFT push button again. It probably appears to you that the Y020 output has remained on, and that the shift register simply energized the Y021 output. However, two things happened when you pressed the X000 SHIFT push button the second time. First, the signal at R016 was shifted one position to R017, which turned on Y021. Second, the Y020 was left on because R016 received the new signal through the NC X007 DATA input.

Pressing the X000 SHIFT push button again causes the same thing to happen. Except now the data in R017 is shifted to R018, the data in R016 is shifted to R017, and R016 receives the new signal through the NC X007 DATA input.

Repeatedly pressing the X000 SHIFT push button continues this action. When the last output light has been energized, repeatedly pressing the X000 SHIFT push button does not appear to change anything, yet the shifting continues. However, the data from R01D becomes overflow and is lost, whereas the data from R01C is continuously shifted to R01D with every operation of the X000 SHIFT push button.

Let's examine a change to this shift register circuit. Reset the circuit back to the original condition by pressing the X001 RESET push button. All output lights that were on will now turn off. Rotate the X007 DATA selector switch to the second position. This causes the NC contact of X007 to change state and become open. This means that the DATA signal is a zero.

Now when you press the X000 SHIFT push button, nothing happens, because the X007 DATA signal is zero. Repeatedly pressing the X000 SHIFT push button does not cause any of the outputs to be energized.

Now I realize that at no time during these two tests have you been able to see a shifting of data. And I don't really expect you to take my word for it. So, one last demonstration will show some shifting of data. Reset the circuit back to the original condition by pressing the X001 RESET push button. Rotate the X007 DATA selector switch back to the original position.

Press the X000 SHIFT push button, and output Y020 is energized. Rotate the X007 DATA selector switch to the second position, and press the X000 SHIFT push button again. The shift has been put into effect, and now you can see it. You can see that the output from Y020 has been shifted to Y021, and the zero (0) input from the X007 DATA input to R016 has shut off output Y020.

By alternating the position of the X007 DATA input selector switch, and then pressing the X000 SHIFT input push button after each alternation, you will observe the shifting of data from one output light to the next.

There are even more dramatic ways to see this shifting effect. First press the X001 RESET push button to reset the shift register. Rotate the X007 DATA selector switch and press the X000 SHIFT push button to get output Y020 to energize. Rotate the X007 DATA selector switch again and press the X000 SHIFT push button several times. You will now see a single output light moving across the PLC panel from Y020 to Y027.

For further experimentation with this circuit, you can modify the pattern of operating the X007 DATA and X000 SHIFT inputs to see other patterns of output lights being shifted across the PLC panel.

This concludes the presentation on the basics of shift registers. Another shift register circuit is presented in chapter 5 in the section entitled "Lead-Lag Circuits."

4 THE PRACTICE CIRCUIT

GETTING STARTED PROGRAMMING

This chapter shows you how to construct the most basic type of circuit. The section entitled "Synonyms and Descriptions" in chapter 2 led you through the sequence of steps for establishing a name for your program, and how to get to the point at which you can start to write a program. So let's do it now.

Start up the EX-PDD program, then proceed as follows:

1. From the TOP MENU, select F7–Enter/Change program name.
2. While at the Enter/Change program name screen, type over the word *default* with the program name of your choice. Establish a program name of your own (other than PRACTICE). Note that your program names must be written in the same manner as in DOS. The program name cannot contain any spaces, and you are limited to a name that is only eight characters long.
3. Press F10. This saves the program name and brings you back to the TOP MENU.
4. Press F2–Ladder Logic.

You now have a blank screen, ready for your programming efforts. From command line 1, press F4–Edit Page. This brings up command line 2. Now press F1–Seq. Inst. and start to work with the following circuits. Remember to press Shift-F10 to ACCEPT (to SAVE) each page when you have finished it. When you are ready to program the next page, simply press F2–New Page. Now you are ready to start programming.

The first circuit contains only two devices, a START push button, represented by a normally open (NO) contact, and an output VALVE. When you open a new page and begin editing or programming, the cursor appears in the upper left-hand corner of the screen. Make sure that you are at command line 3 and press F1--] [-. Then move the cursor over one

Figure 4.1 Rung

```
|
|-| |---------------------------------------------------------( )-
|
```

Figure 4.2 Rung after Assignment of Addresses

```
|
|-| |---------------------------------------------------------( )-
|X000                                                      Y020
```

space and press F5 -( ). The coil device symbol automatically is placed at the right-hand side of the screen and the rung looks like Figure 4.1.

Warning: After you have entered your input, do not use F4 to create a series of short lines that complete the rung over to the coil position and then insert the coil. The rung will look fine, and it will work, but this method uses more memory steps than the procedure described in the previous paragraph. You should avoid using more memory than needed. The more memory steps used, the longer the scan time of the programmable logic controller (PLC).

Assign the address of Y020 to the output. Remember to press ENTER before leaving this location. Then move the cursor to the input and assign the address X000 to the input. The rung should now look like Figure 4.2.

Review the section entitled "Synonyms and Descriptions" in chapter 2, if necessary, and assign the synonym START to X000 and the synonym VALVE to Y020. When you exit that operation, the rung will look like Figure 4.3.

This is the most basic type of circuit that you can develop. All it does is energize the VALVE when you press the START push button. The VALVE stays on only for as long as you press the START push button. Whenever you release the START push button, the VALVE shuts off.

To turn this into a holding circuit, place the cursor on the X000 input and press F3–Close. This inserts a short vertical line to the right of, and

Figure 4.3 Rung after Assignment of Synonyms

```
|START                                                     VALVE
|-| |---------------------------------------------------------( )-
|X000                                                      Y020
```

Figure 4.4 Rung with a Holding Circuit Contact

```
|START                                                        VALVE
|-| |--+------------------------------------------------------( )-
|X000  |                                                      Y020
|      |
|VALVE |
|-| |--+
|Y020
```

Figure 4.5 Rung with a Holding Circuit Contact (Screen Image)

```
|START                                                        VALVE
|-| |--+------------------------------------------------------( )-
|X000  |                                                      Y020
|VALVE |
|-| |--+
|Y020
```

drops down from, the original position. Move the cursor down one rung, insert a NO contact, and give it the address of Y020. The circuit looks like Figure 4.4.

Remember that throughout this book an extra line has been inserted in each circuit between each rung. This has been done to make the diagram easier to read, especially if you are copying the circuit to your personal computer (PC) screen. The screen itself will look like Figure 4.5.

On your machine or system, you would wire the START push button to the X000 input terminal of the PLC and wire the VALVE to the Y020 output terminal of the PLC. Now it's time to do some experimenting.

Try each key on command line 3 to see the results. You can't cause any problems or do any damage, so try them all.

Then try changing a NO contact to a normally closed (NC) contact. Try erasing a device. Leaf through the pages of this book and pick out some circuits to copy. The more time you spend doing this, the greater the level of confidence you will have about using the PLC and the EX-PDD software.

PRACTICE CIRCUIT

This chapter provides you with the opportunity to practice the complete process of entering a working circuit into the EX-PDD250 software. This and the next five sections take you through the process of creating a circuit on your PC, loading the circuit into your PLC, operating the PLC, performing on-line changes, and saving the modified circuit from the PLC back to the PC.

You need to gain some experience with the system so that you can begin to feel comfortable while working with the EX-PDD software. You can do that by copying the circuit in this section.

The input and the output devices in this circuit have been set up so that the circuit can be programmed into all of the PLCs in the medium series. If you are working with an M20 or an M40 PLC, just start right in. If you are working with an EX100 PLC, install a digital input module into slot 1 and a digital output module into slot 3. If you are working with an EX250 or an EX500 PLC, install a digital input module into slot 0 and a digital output module into slot 2.

Start the EX-PDD program, then proceed as follows:

1. From the TOP MENU, select F7–Enter/Change program name.
2. While at the Enter/Change program name screen, type over the word *default* with the word **PRACTICE**, or with a program name of your choice.
3. Press F10. This saves that name and returns you to the TOP MENU.

Before you start to enter the circuit, take a few minutes now (if you didn't do it while you were reading the section entitled "Synonyms and Descriptions" in chapter 2) to enter the *device commentary synonym* and the *device commentary description* notes into the program.

If you have just completed entering the synonyms and descriptions, you should be at command line 1. If not, you need to get there. You now have a blank screen, ready for your programming efforts.

From command line 1, press F4–Edit Page. This brings up command line 2. Now press F1–Seq. Inst. and start to copy the following circuit. Remember to press Shift-F10 to ACCEPT (to SAVE) each page when you have finished it. When you are ready to program the next page, simply press F2–New Page. Figures 4.6 and 4.7 show the Practice Circuit.

Figure 4.6 Practice Circuit (Page 1)

```
|PBSTRT COUNT                                                      START
|-| |--+-|/|-----------------------------------------------------------( )-
|X001  |R002                                                       R000
|      |
|START |
|-| |--+
|R000
|
|START  RUN                                                        ALARM
|-| |--+-|/|-----------------------------------------------------------( )-
|R000  |R001                                                       Y020
|      |
|      |                          TIM006                           RUN
|      +[00030   TON    T000 ]--|/|--------------------------------( )-
|                                 R00F                             R001
|
|RUN                                                               TIM001
|-| |--+[00020   TON    T001 ]-------------------------------------( )-
|R001  |                                                           R00A
|      |
|      |                                                           TIM002
|      +[00040   TON    T002 ]-------------------------------------( )-
|      |                                                           R00B
|      |
|      |                                                           TIM003
|      +[00060   TON    T003 ]-------------------------------------( )-
|      |                                                           R00C
|      |
|      |                                                           TIM004
|      +[00080   TON    T004 ]-------------------------------------( )-
|      |                                                           R00D
|      |
|      |                                                           TIM005
|      +[00100   TON    T005 ]-------------------------------------( )-
|      |                                                           R00E
|      |
|      |                                                           TIM006
|      +[00120   TON    T006 ]-------------------------------------( )-
|                                                                  R00F
|
|TIM006 +-----------+                                              COUNT
|-| |---C  CNT      Q----------------------------------------------( )-
|R00F   |           |                                              R002
|       |           |
|START  |           |
|-| |---E 00005 C00 |
|R000   +-----------+
```

Figure 4.7 Practice Circuit (Page 2)

```
|TIM001 TIM006                                                     LIGHT1
|-| |----|/|-------------------------------------------------------( )-
|R00A   R00F                                                       Y021
|
|TIM002 TIM005                                                     LIGHT2
|-| |----|/|-------------------------------------------------------( )-
|R00B   R00E                                                       Y022
|
|TIM003 TIM004                                                     LIGHT3
|-| |----|/|-------------------------------------------------------( )-
|R00C   R00D                                                       Y023
|
| +----+
|-| END|
| +----+
```

VALIDATE

To prepare for going on-line, send the circuit through the VALIDATE routine, which was described in chapter 2 in the section entitled "EX-PDD Software Command Line Menus." Performing the VALIDATE routine now may save you some headaches later when you attempt to load your circuit into the PLC. The VALIDATE routine checks the circuit for gross errors only. It checks for devices that do not have addresses, for openings in rungs, for short circuits, for vertical lines that are not connected, for pages that contain too many instructions, and for other errors that the software does not allow you to enter into the PLC.

VALIDATE will *not* tell you if your logic is correct. It does not tell you if your circuit will operate as you intended to program it. It does not tell you that the timer you wanted set for thirty seconds will time out at three seconds. Nor does it tell you that you have programmed your largest motor to be cycling on and off every two seconds.

However, in the case of the EX-PDD250 software, you cannot perform the VALIDATE routine while you are in the off-line mode until after you have manually performed an Input-Output (I/O) Allocation. To start the manual I/O Allocation operation, go to the TOP MENU and press F1–System and Control Information. Press F1–System Information and type **PRACTICE** (or your program name) on the first line. Move the cursor down to line six and press the space bar until your type of PLC is displayed. Press F10 to save this information and return to the previous screen.

Press F2–I/O Allocation. The new screen allows you to specify which type of module you have installed in each slot. Where to install each module and which registers and addresses are assigned to these modules are covered in chapter 6 in the section entitled "I/O Allocation."

For the "Practice Circuit" on which you are now working, move the cursor to the first open slot and press F1–Input X. Leave the second slot blank, and move the cursor down to the third slot. Press F3–Output Y. Press F10 twice to save this information and to return to the TOP MENU.

You can elect to skip the VALIDATE routine if you do not want to perform the manual I/O Allocation. In that case, proceed to the next section, entitled "Going On-line," which leads to the section entitled "Automatic Identification of the I/O Allocation."

When you start to load your circuit into the PLC, the LOAD operation will be stopped if your circuit contains an error that would have been found if you had performed the VALIDATE routine. A message will be displayed on your screen indicating that you have an error and on which page the error occurs. These errors must be corrected one at a time before you can load your circuit into the PLC.

Return to your circuit. To perform the VALIDATE on any page, press F7–Valid Ladder at command line 1. When you press Shift-F7–Valid Ladder, the program checks each page sequentially and then (if all is well) gives you a message that reads as follows:

```
Validating Ladder Logic Complete. 54 Steps Used.
```

If you have made an error that VALIDATE does not accept, the validation process is stopped, and the program takes you to the page that contains the first problem. It puts the program into the EDIT mode and places the cursor at the point of the error. All you have to do is recognize the nature of the error and correct it. Make the correction, ACCEPT the page, and start the Shift-F7–Valid Ladder routine again.

Repeat this process until you get the following message:

```
Validating Ladder Logic Complete. 54 Steps Used.
```

GOING ON-LINE

At this stage, you should have a working and validated circuit to load into your PLC. If you haven't already done so, run power to the PLC and hard-wire a push button to input X001.

The Sentinel Key and Converter Module must now be installed. Figure 4.8 shows the M20 PLC, the Sentinel Key, and the Converter Module to assist you with the rest of the setup.

The Sentinel Key is the small device shown at the lower right edge of the M20 PLC. It is packaged with the EX-PDD250 software. Install the Sentinel Key into the proper port of your PC: the *serial* port for Version 3.X or the *parallel* port for Version 2.X.

Note: The Sentinel Key became a part of the EX-PDD250 software with Version 2.40. The Sentinel Key allows the EX-PDD software to be installed in any number of PCs, unlike the previous restriction of two active installations for the original EX-PDD.

For a PC to communicate with a PLC, the Sentinel Key must be installed on that particular computer.

Install the Converter Module (Part no. EX25GP232A-TIC2). This module is the black box shown at the lower left edge of the M20 PLC in Figure 4.8 and comes with its own connectors. Plug one end into the serial

Figure 4.8 M20 with Accessories

port (preferably the COM1 port) of the PC, and plug the other end into the programming port of the PLC.

Return to the TOP MENU screen. Press F6–Offline/Online programming. If all connections are correct, you will receive a message that reads as follows:

```
Busy Initializing Communications.
```

The process takes less than five seconds, and then the line just above the command line changes from off-line status to on-line status and reads as follows:

```
Online Unit: S Halt EX100 1022 ID: PRACTICE Pages: 1 Steps: 1
```

If you experience a problem going on-line, there are seven possible reasons, which can be classified into two groups.

The first group of problems has to do with the Sentinel Key, and you will immediately receive a message that reads as follows:

`Hardware key not installed.`

This message appears for any of the following three reasons:

1. You may not have installed the Sentinel Key.
2. You may not have installed the Sentinel Key in the correct port of your PC.
3. The controller card in your PC that contains the connectors for the serial ports (normally called COM1 and COM2) and the parallel port (normally called LPT1) may not be compatible with the Sentinel Key. If you have installed the Sentinel Key correctly and you continue to receive the "Hardware key not installed" message, you must change the controller card. Do not replace it with a new card of the same model. The original card is probably not defective; the problem is that the card is not compatible with the Sentinel Key. Try a card from another manufacturer.

The second group of problems involves the Converter Module or the cable. After about forty-five seconds you will receive the following message:

`Not connected, check cables/Computer Link Card for proper installation.`

This second message can appear for one of the following four reasons:

1. You may not have installed the Converter Module or not connected both ends of the cable.
2. You may not have installed the Converter Module into a serial port of your PC.
3. You may have plugged the Converter Module into the COM2 port of your PC. If you have connectors or gender changer connectors available and can plug the Converter Module into the COM1 port, do it now. If not, press F8–Enter/Change Communication parameters. Press F4–COM2; then press F10 to save this information. This step is required every time you start your computer with the intention of communicating with your PLC.
4. If you are using the Computer Link Card, you may not have installed it correctly. Refer to your Link Card manual for assistance in this situation.

AUTOMATIC IDENTIFICATION OF THE I/O ALLOCATION

If you did not perform the manual I/O Allocation operation described in the section entitled "Validate," now is the time to perform the automatic identification of the I/O Allocation.

The automatic method of I/O Allocation allows the PC to communicate with the PLC. It checks to see which type of I/O module has been installed in each slot of the PLC. It also assigns the registers to each of the modules. Therefore, the automatic method can be used only when you have the actual PLC that you will be using in your operation and you are ready to load the program into it.

This automatic procedure simplifies start-up, but you still need to understand I/O Allocation and the register assignment operation before you proceed with programming. For a more complete discussion about the subject of I/O Allocation, including the manual method of performing an I/O Allocation, refer to the section entitled "I/O Allocation" in chapter 6.

To use the automatic identification feature of any of the Medium PLCs, you need to set up the PLC as it is to be used in your operation. This must include either the expansion module or the expansion rack with the I/O modules installed, if either one of these expansion features is to be used. This initial setup includes the connections for the PLC power and the connections to your PC, which includes the Sentinel Key and the Converter Module.

Move the Program/Link switch to the Program position. On the M20 and the M40 PLCs, this is dual-in-line package (DIP) switch 5, which is located behind the cover that contains the status light-emitting diodes (LEDs) and the RUN-P/RUN/HALT switch. On the EX100, the EX250, and the EX500 PLCs, this is a slide switch on the front face of the central processing unit (CPU).

Move the RUN-P/RUN/HALT switch to the HALT position. Starting at the TOP MENU, use the following list to perform the Automatic Allocation:

1. Press F7–Enter/Change program name. Overwrite the word *default* with the word **PRACTICE** or your program name, and press F10 to save that name and to exit the function.
2. If you have plugged the Converter Module into the COM2 port of your PC, press F8–Enter/Change Communication parameters, press F4–COM2, and then press F10 to save the information and return to the TOP MENU.

3. Press F1–System and Control Information. Then press F1–System Information and type **PRACTICE** or your program name on the first line. Move the cursor down to line six, and press the space bar until your type of PLC is displayed. Press F10 twice to save this information and to return to the TOP MENU.
4. Press F6–Offline/Online programming selection to go on-line. When you are taken on-line, the *off-line* comment at the left-hand side of the command line at the bottom of your screen is changed to *on-line*.

   Note: If your PLC is not allowed to go on-line, you will receive one of two messages. The first one reads as follows:

   `Hardware key not installed.`

   The second message reads as follows:

   `Not connected, check cables/Computer Link Card for proper installation.`

   If you receive either of these messages, read the section entitled Going On-line for the possible remedies.
5. Once you are on-line, press F1–System and Control Information. This brings up a new menu.
6. Press F2–I/O Allocation. This brings up another new menu.
7. Press F8–Auto Setup. You will receive a message that reads as follows:

   `Auto Setup complete.`

The automatic identification of the I/O Allocation is complete. To save this information and to return to the TOP MENU, press F10 twice.

TESTING THE PRACTICE CIRCUIT

The next step in the process is to take the circuit you have designed in your PC and load it into your PLC.

LOAD

Start at the TOP MENU, and proceed as follows:

1. Press F1–System and Control Information.
2. At the next screen, press F3–Load/Save/Compare.
3. At the next screen, press F1–Load Ladder.
4. Use the HALT/RUN switch on the PLC to put the PLC in HALT, and press F9 to start the LOAD.
5. When the operation is complete, the message reads as follows:

Load Complete

6. Pressing F10 three times returns you to the TOP MENU.

TEST

To test the circuit, move the HALT/RUN switch on the PLC to the RUN mode, and press the X001 PBSTRT push button. If you have entered the program correctly, the following is a description of how the input and outputs operate, and how the circuit should function:

1. The PBSTRT (X001) push button starts the sequence.
2. The ALARM (Y020) starts immediately, runs for 3.0 seconds (Timer T000), and shuts off.
3. A delay of 2.0 seconds (T001) occurs, and LIGHT1 (Y021) turns on. After an accumulated delay of 4.0 seconds (T002), LIGHT2 (Y022) turns on. After an accumulated delay of 6.0 seconds (T003), LIGHT3 (Y023) turns on.
4. The three output lights then turn off in the reverse sequence at 2.0-second intervals (T004, T005, and T006).
5. The sequence of the three output lights turning on and off repeats itself five times (C00), and then the whole circuit shuts itself off.

If the circuit is not performing as expected, compare each device and address in the circuit on the screen with its circuit in the section entitled "Practice Circuit." One problem may be that the EX-PDD250 software has loaded your circuit into the PLC with some or all of the inputs in a forced-off mode. If this has happened, when you press the X001 PBSTRT push button, the circuit will not operate. If this is the case, the address for the push button will appear on the screen as *x001,* instead of as *X001*.

To correct this problem, use the following steps:

1. Go to the TOP MENU, and press F1–System and Control Information.
2. Press F1–System Information.
3. Press F1–Reset Commands.
4. Press F6–Force Clear. You will receive the following message:

   Press [Enter] to confirm Force Clear, any other key to cancel.
5. Press ENTER, and the following message appears:

   Force Clear command complete.
6. Press F10 two times, and you will be returned to the TOP MENU.

If your circuit is working as described, you can examine an excellent troubleshooting feature in the EX-PDD software.

Leave or put the PLC in the RUN mode. From the TOP MENU, press F2–Ladder Logic to bring the relay ladder logic (RLL) back to the screen. At the new command line, press F9–Monitor Force. Some of the lines on the screen will become brighter. These brighter lines indicate that current is flowing in those rungs of the circuit. Generally speaking, these brighter lines usually cannot be readily distinguished from the regular lines. However, notice that the F9 command has changed to F9–Change Color.

By pressing F9 again, you see a much more noticeable change. If your monitor screen is monochrome, the powered lines begin flashing. If you have a color monitor, the powered lines change to a color that is different from the unpowered lines and they begin flashing. This means that you have a powerful debugging tool at your disposal. How can this feature help you?

You can view a page of your circuit and see which rungs are operating at any given time. This screen image can be compared with the output lights on the PLC. If the rung on the screen is flashing, but the output light is not on, there is a problem in the program. It could mean that there is an error in the way you programmed the output relative to a timer.

If the rung is flashing and the output light is on, but the output device on your machine is not operating, the output device may be wired to the wrong PLC output terminal. It could also mean that the wiring between the output terminal, the actual output, and the power source is the problem.

This feature is especially valuable when you work with a circuit that uses quite a few internally generated PLC relays. There is no other way to see if these relays are working correctly if the portion of the system that you are debugging is not working.

When you want to go off-line, leave the circuit, and shut down the PLC, you need to perform one more operation. This involves saving the circuit to the EEPROM.

Save the Program to the EEPROM

Medium PLCs use two methods to store data—random-access memory (RAM) and electrically erasable programmable read-only memory (EEPROM). The programmed circuit is automatically loaded into RAM at start-up, and actually runs from the RAM. However, the RAM cannot retain the program when the PLC is turned off, whereas the EEPROM can. Therefore, the program must be written to the EEPROM before you exit the PLC.

To save a program to the EEPROM, follow these steps:

1. Put the PLC into the HALT mode.
2. Go to the TOP MENU, and press F1–System and Control Information.
3. Press F1–System Information.
4. Press F5–Write EEPROM. Nothing happens, however, until the EEPROM has been saved. You will then receive a message that reads as follows:

   `Write EEPROM complete.`

5. Press F10 two times, and you are brought to the TOP MENU.
6. Exit the EX-PDD250 software and shut down the PLC.

EXPERIMENTING WITH THE PRACTICE CIRCUIT

Now it's time to do some experimenting. You now have a working circuit. This section describes how easy it is to modify the circuit and shows how easy it is to modify the preset values in the timers and counters.

Note: Editing in the on-line mode is the same as editing in the off-line mode. Changes can be made to the circuit with the PLC in either the RUN mode or the HALT mode.

To start, let's make some changes to the timers. The initial flashing rate of the outputs is 2.0 seconds, beginning after the ALARM delay. Leave T000 as it is, but change the flashing rate of the outputs to 1.0-second intervals.

When making these changes, start with timer T001. Insert a 1.0-second value by typing **00010** over the original time value in the timer; be sure to include all the zeros. Remember to press ENTER before moving to the next timer. Then change the values of timers T002, T003, T004, T005, and T006 to **00020**, **00030**, **00040**, **00050**, and **00060** seconds.

An ACCEPT (Shift-F10) of the page must be done to save these changes. If you make these timer changes while the PLC is in the RUN mode, you will receive a message that reads as follows:

`WARNING! Program Change in RUN Mode. [Enter] to Confirm, [Esc] to Abort.`

Press ENTER, and you will receive the following message:

`Saving Page.`

Start the circuit by pressing the X001 PBSTRT push button to see if you have been successful with the changes.

After you have completed the timer modifications, return to the program to change the count value of the counter from five cycles to ten cycles of the flashing sequence; then test it.

One more change. Change the sequence of the flashing lights so that the outputs shut off in the same order that they come on. To do this, move to page 2 of the circuit.

On rung number one, change the address of the NC R00F contact to **R00D**. On rung number three, change the address of the NC R00D contact to **R00F**. That's all there is to it. Accept these changes and test it out.

You have successfully demonstrated to yourself that PLC programs can be easily changed, even on-line and in the RUN mode. However, you also need to realize that *the program now in the PLC is no longer the same as the program you first wrote, which is the program still in your PC.* You must perform a SAVE operation, which overwrites the original program in the PC with the latest version of the program in the PLC.

SAVE

To save the RLL in the PLC back to the hard disk of the PC, start at the TOP MENU, and proceed as follows:

1. Press F1–System and Control Information.
2. At the next screen, press F3–Load/Save/Compare.
3. At the next screen, press F2–Save Ladder.
4. The new screen lists three options under the following title:

   `Select PLC areas:`

   For this simple circuit, all that must be saved is the ladder logic portion.
5. Put the PLC in the HALT mode, and press F9 to start the SAVE operation.
6. You will receive the following message:

   `Busy save PLC in progress.`

   When the operation is complete, the message reads as follows:

   `Save Complete.`
7. Pressing F10 three times returns you to the TOP MENU.

Save the Program to the EEPROM

The final step is to save your program to the EEPROM. To do so, use the following steps:

1. Put the PLC into the HALT mode.
2. Go to the TOP MENU, and press F1–System and Control Information.
3. Press F1–System Information.
4. Press F5–Write EEPROM. After the EEPROM has been saved, you will receive the following message:

   `Write EEPROM complete.`

5. Press F10 two times, and you are brought to the TOP MENU.
6. Exit the EX-PDD250 software and shut down the PLC.

This completes the experimental phase of this book. I strongly recommend that you use this circuit for more programming efforts and that you create your own new circuits to gain additional experience with your PLC.

I also recommend that as you work on programs for your equipment, that you have a spare PLC available and that it be wired with several input push buttons and a two-position or a three-position selector switch. It is always a good idea to take a section that you have just programmed, load it into the PLC, and verify that the circuit is operating the way you designed it.

5 OPERATIONAL CIRCUITS

HOLDING CIRCUITS

Although a holding circuit has only one purpose, it can be used in many ways. The purpose of a holding circuit is to allow an input signal to be activated for a short period of time, then be released, while the output stays on.

Two of the many uses of a holding circuit are as follows:

- With a push button, a motor or a fan continues to run even after the START push button has been released.
- With a limit switch, the signal can be held after a shipping case has activated and then released a limit switch so that the shipping case can be diverted at the next station.

In its simplest form, a holding circuit looks like Figure 5.1.

When the START push button is pressed, the normally open (NO) contact is closed. This action provides the power to energize VALVE1, which you might be using to allow city water to flow to a holding tank. In this case, Y021 is also used as an input. It is placed parallel with the X001 NO contact, and it automatically closes when output Y021 is ener-

Figure 5.1 Holding Circuit—Start Only

```
|START                                                        VALVE1
|-| |--+------------------------------------------------------( )-
|X001  |                                                      Y021
|      |
|VALVE1|
|-| |--+
|Y021
|
| +----+
|-| END|
| +----+
```

gized. The closing of input Y021 allows current to flow through inputs X001 and Y021 to output Y021.

If you release START push button X001, a path still exists for current to flow through input Y021 to output Y021, thus holding output VALVE1 on, even though the START push button has been released and that contact is again open.

This circuit has no value of its own other than to demonstrate the concept of a holding circuit. There is no way to shut this circuit off, other than to pull the plug.

The next circuit adds a STOP push button (Figure 5.2).

The "Holding Circuit-Start and Stop" is a very common circuit, and you will find that it has many uses. It operates the same as a start-only holding circuit, except that the output can be shut off when someone presses and releases the STOP push button. Note that the STOP push button is shown as normally closed (NC).

In the section entitled "Inputs" in chapter 3, I recommended that all inputs be *hard-wired* NO. This example follows that recommendation. The STOP push button is *programmed* NC. This means that current can flow through X002 whenever X001 is pressed. It also means the circuit can be stopped if someone presses X002. This shuts off all current flow, whether the output is being energized through input X001 or through input Y021.

A modified start-and-stop holding circuit can be programmed that uses an internal programmable logic controller (PLC) relay (Figure 5.3).

If you were to look at the operation of this circuit in real life, you would not see any difference between it and the "Holding Circuit-Start and Stop." The difference is that an internal relay R000 has been programmed into the circuit and is used here for two separate purposes. First, R000 is used as the holding contact around X001. Second, it is the contact used to allow power to flow to output Y021.

Figure 5.2 Holding Circuit—Start and Stop

```
|START   STOP                                                    VALVE1
|-| |--+-|/|-------------------------------------------------------( )-
|X001   |X002                                                    Y021
|       |
|VALVE1 |
|-| |--+
|Y021
|
| +----+
|-| END|
| +----+
```

Figure 5.3 Holding Circuit—Start and Stop: PLC Relay-Operated

```
|START   STOP                                                 RUN
|-| |--+-|/|--------------------------------------------------( )-
|X001  |X002                                                  R000
|      |
|RUN   |
|-| |--+
|R000
|
|RUN                                                          VALVE1
|-| |---------------------------------------------------------( )-
|R000                                                         Y021
|
| +----+
|-| END|
| +----+
```

The advantage of this method is that it establishes a RUN function relay that can be used an unlimited number of times in the program, regardless of what happens later to VALVE1.

Holding Circuit—Timed Stop

Let's look at a way to start VALVE1 and to stop it after a specified length of time. This circuit adds an automatic shutoff of VALVE1 after 120.0 seconds while still allowing the operator to use the STOP push button to shut off the valve within that time span if necessary.

Timer T017 and output VALVE1 Y021 are energized together by NO contacts of the RUN relay R000 when the circuit is started by pressing the START push button. After 120.0 seconds, the timer times out and energizes the R01F T-STOP relay. This causes the NC contact of R01F on the first rung of the circuit to open. This action causes the RUN relay to be de-energized, shuts off VALVE1 and the timer, and sets up the circuit for another start-up (Figure 5.4).

Holding Circuit—Start and Second Stop: PLC Relay-Operated

Let's take a look at a circuit that utilizes the concept of a PLC relay as the holding relay and combine it with a second means of shutting off VALVE1.

Imagine that this program goes on for many pages, and another relay UNFILL (R012) is energized that determines when to shut off VALVE1. An NC contact of internal relay R012 is used in the rung with the Y021

Figure 5.4 Holding Circuit—Timed Stop

```
|START   STOP    T-STOP                                          RUN
|-| |--+-|/|----|/|---------------------------------------------( )-
|X001   |X002    R01F                                            R000
|       |
|RUN    |
|-| |--+
|R000
|
|RUN                                                             T-STOP
|-| |---[01200   TON    T017 ]----------------------------------( )-
|R000                                                            R01F
|
|RUN                                                             VALVE1
|-| |-----------------------------------------------------------( )-
|R000                                                            Y021
|
| +----+
|-| END|
| +----+
```

output. There are now two ways to shut off VALVE1. The valve can be shut down automatically by R012 UNFILL as the rest of the program continues, or the valve and the complete program can be shut off when someone presses the STOP push button (Figure 5.5).

As you can see, many variations of holding circuits are possible.

Any circuit in this book that shows a valve as the output can also be used for a fan, an electromechanical relay, a motor starter, a pilot light, or any other type of output. More examples of holding circuits are discussed in the next section.

Figure 5.5 Holding Circuit—Start and Second Stop: PLC Relay-Operated

```
|START   STOP                                                    RUN
|-| |--+-|/|----------------------------------------------------( )-
|X001   |X002                                                    R000
|       |
|RUN    |
|-| |--+
|R000
|
|RUN     UNFILL                                                  VALVE1
|-| |----|/|----------------------------------------------------( )-
|R000    R012                                                    Y021
|
| +----+
|-| END|
| +----+
```

MOTOR STARTER CIRCUITS

There are a number of ways to control the starting of a motor, and only a few are discussed here. Let's start by assuming that your motor will be controlled by a motor starter and that the motor starter will be connected directly to the PLC because the motor starter coil has a lower current draw than the rating of your PLC output.

Jog

In the jog system situation, the motor starter M1STRT is turned on for as long as the JOG push button is held down (Figure 5.6).

Automatic

In the automatic start system, the motor starter M1STRT is turned on when the START push button is pressed momentarily, and the motor runs until the STOP push button is pressed momentarily. The PLC-generated Y021 input is used as a holding contact around the START push button, which allows the operator to release the START push button while the motor continues to run. This PLC-generated Y021 input is used as a holding contact around the START push button in the same way as an auxiliary contact of a motor starter is used in a hard-wired version of this circuit (Figure 5.7).

Automatic with Jog

In a combination of the "Jog" and "Automatic" systems, the motor starter M1STRT can be turned on in either of the two ways. Because of the dual function of this circuit the "Automatic" circuit just discussed must be modified.

Figure 5.6 Jog

```
|JOG                                                          M1STRT
|-| |------------------------------------------------------------( )-
|X000                                                          Y021
|
| +----+
|-| END|
| +----+
```

Figure 5.7 Automatic

```
|START   STOP                                                   M1STRT
|-| |--+-|/|--------------------------------------------------------( )-
|X001  |X002                                                    Y021
|      |
|M1STRT|
|-| |--+
|Y021
|
| +----+
|-| END|
| +----+
```

The first way to run the motor is to press the START push button. In this case, an internal PLC RUN relay (R000) was created to act as the holding contact around the START push button, which allows the operator to release the START push button. The program also isolates the M1STRT output by placing it in a new rung with a NO R000 RUN contact to turn on M1STRT directly. Again, the motor continues to run until the STOP push button is pressed.

The second way to run the motor is to press the JOG push button. In this case, M1STRT runs the motor for only as long as the JOG push button is held down (Figure 5.8).

Figure 5.8 Automatic with Jog

```
|START   STOP                                                   RUN
|-| |--+-|/|--------------------------------------------------------( )-
|X001  |X002                                                    R000
|      |
|RUN   |
|-| |--+
|R000
|
|RUN                                                            M1STRT
|-| |--+------------------------------------------------------------( )-
|R000  |                                                        Y021
|      |
|JOG   |
|-| |--+
|X000
|
| +----+
|-| END|
| +----+
```

Automatic–Timed On

The "Automatic–Timed On" start system is a variation of the "Automatic" Circuit that provides a safety feature during the starting of a motor or any other system. In this situation, the motor starter M1STRT is turned on after a thirty-second delay from the time when the START push button is pressed. Then the motor runs until the STOP push button is pressed.

The R000 relay serves as both the holding contact around the START push button and as the NO contact that starts the timer (T000). At the end of the thirty-second delay, M1STRT is energized (Figure 5.9).

Automatic Sequence

Method 1

You may want an automatic sequence to start several motors in a specific sequence with timed intervals between the starts, but you only want to press the START push button once. There are two different ways to handle time intervals between motor starts. The choice lies in the method of describing the timing sequence.

Let's use a three-motor system as an example. In Method 1, the time of reference is time zero (0). You want the first motor to start immediately, the second motor to start at an elapsed time of forty-five seconds, and the third motor to start at an elapsed time of ninety seconds (Figure 5.10).

The preceding circuit can be simplified somewhat (Figure 5.11).

Figure 5.9 Automatic–Timed On

```
|START   STOP                                                  RUN
|-| |--+-|/|---------------------------------------------------( )-
|X001   |X002                                                  R000
|       |
|RUN    |
|-| |--+
|R000
|
|RUN                                                           M1STRT
|-| |---[00300  TON   T013 ]-----------------------------------( )-
|R000                                                          Y021
|
|  +----+
|-| END|
|  +----+
```

Figure 5.10 Automatic Sequence—Method 1

```
|START  STOP                                                      RUN
|-| |--+-|/|-----------------------------------------------------( )-
|X001   |X002                                                    R000
|       |
|RUN    |
|-| |--+
|R000
|
|RUN                                                             M2TIME
|-| |---[00450   TON    T002 ]-----------------------------------( )-
|R000                                                            R00A
|
|RUN                                                             M3TIME
|-| |---[00900   TON    T003 ]-----------------------------------( )-
|R000                                                            R00B
|
|RUN                                                             M1STRT
|-| |------------------------------------------------------------( )-
|R000                                                            Y021
|
|M2TIME                                                          M2STRT
|-| |------------------------------------------------------------( )-
|R00A                                                            Y022
|
|M3TIME                                                          M3STRT
|-| |------------------------------------------------------------( )-
|R00B                                                            Y023
|
| +----+
|-| END|
| +----+
```

Figure 5.11 Automatic Sequence—Method 1 (Simplified)

```
|START  STOP                                                      RUN
|-| |--+-|/|-----------------------------------------------------( )-
|X001   |X002                                                    R000
|       |
|RUN    |
|-| |--+
|R000
|
|RUN                                                             M2TIME
|-| |--+[00450   TON    T002 ]-----------------------------------( )-
|R000   |                                                        R00A
|       |
|       |                                                        M3TIME
|       +[00900   TON    T003 ]-----------------------------------( )-
|       |                                                        R00B
|       |
|       |                                                        M1STRT
|       +--------------------------------------------------------( )-
|                                                                Y021
|
|M2TIME                                                          M2STRT
|-| |------------------------------------------------------------( )-
|R00A                                                            Y022
|
|M3TIME                                                          M3STRT
|-| |------------------------------------------------------------( )-
|R00B                                                            Y023
|
| +----+
|-| END|
| +----+
```

Method 2

In Method 2, the time of reference is defined as the time lag from the start of one motor to the start of the next motor. In this case you decide you want the first motor to start immediately, the second motor to start forty-five seconds after the first has started, and the third motor to start forty-five seconds after the second motor has started (Figure 5.12).

The method chosen depends on the way that you describe the sequence or on the way that you think changes might be made once the system is in operation. All of the circuits presented result in the same operation, even though there are slight differences in design.

FLASHING LIGHT CIRCUITS

This section describes the easiest way to program a flashing light or a continuous on/off device. Toshiba provides eight internally programmed timing relays in the Medium PLCs that are extremely easy to use. They are as follows:

R620 = 0.1 second	R624 = 1 second
R621 = 0.2 second	R625 = 2 seconds
R622 = 0.4 second	R626 = 4 seconds
R623 = 0.8 second	R627 = 8 seconds

These eight timing relays provide a continuous on/off sequence in which the on and off times are equal to one-half of the full time span for that particular timing relay. Therefore, using timing relay R620 provides a continuous timing feature where the on time is 0.05 seconds and the off time is 0.05 seconds. Using timing relay R624 provides a continuous timing feature where the on time is 0.5 second and the off time is 0.5 second.

What is actually provided is the programmed coil portion of the timing relay, and it is always operating. To use it, all you have to do is insert a contact for that timing relay into your circuit. Therefore, not only have you eliminated the programming of the continuous on/off function, but you don't even have to install the coil portion of that timing relay into your circuit.

Figure 5.12 Automatic Sequence—Method 2

```
|START   STOP                                                     RUN
|-| |--+-|/|-----------------------------------------------------( )-
|X001   |X002                                                    R000
|       |
|RUN    |
|-| |--+
|R000
|
|RUN                                                              M1STRT
|-| |------------------------------------------------------------( )-
|R000                                                             Y021
|
|M1STRT                                                           M2TIME
|-| |---[00450   TON    T002 ]-----------------------------------( )-
|Y021                                                             R00A
|
|M2TIME                                                           M2STRT
|-| |------------------------------------------------------------( )-
|R00A                                                             Y022
|
|M2STRT                                                           M3TIME
|-| |---[00450   TON    T003 ]-----------------------------------( )-
|Y022                                                             R00B
|
|M3TIME                                                           M3STRT
|-| |------------------------------------------------------------( )-
|R00B                                                             Y023
|
| +----+
|-| END|
| +----+
```

Consider using this feature on your audible alarms to produce a chirping alarm. It allows you to easily program any output device in a continuously timed on/off mode.

A pair of examples of this idea is shown in Figures 5.13 and 5.14.

Example 1

Example 1 (Figure 5.13) is a simple start/stop circuit in which the gas-on warning pilot light WARNPL (Y021) uses the R624 timing relay and flashes on and off every second as long as the GAS SV (Y020) is operating.

Example 2

Example 2 (Figure 5.14) is a circuit in which the process state pilot light PROCES STATPL (Y023) is on during the normal sequence of operations. During the critical middle portion of the sequence, however, the pilot

Figure 5.13 Flashing Light Circuits (Example 1)

```
|START   STOP                                                  GAS SV
|-| |--+-|/|--------------------------------+------------------( )-
|X000   |X001                               |                  Y020
|       |                                   |
|GAS SV|                                    | FLASHR           WARNPL
|-| |--+                                    +--|/|-------------( )-
|Y020                                          R624            Y021
|
| +----+
|-| END|
| +----+
```

Figure 5.14 Flashing Light Circuits (Example 2)

```
|PROCES                                                      PUMP
|ST/STP                                                      MOTOR
|-| |--+-----------------------------------------------------( )-
|X002  |                                                     Y022
|      |
|      |EMERG.                                               PROCES
|      |SH.DN.                                               RUN
|      +-|/|--+------------------------------------------------( )-
|        R000 |                                              R001
|             |
|             |PROCES                                        PROCES
|             |STAGE1                                        STATPL
|             +-| |---------+----------------------------------( )-
|             |R020         |                                Y023
|             |             |
|             |PROCES       |
|             |STAGE2 FLASHR|
|             +-| |----|/|--+
|             |R040   R624  |
|             |             |
|             |PROCES       |
|             |STAGE3       |
|             +-| |---------+
|              R060
```

light uses the R624 timing relay and flashes on and off every second. During PROCES STAGE1 (R020) and PROCES STAGE3 (R060), the process-status pilot light PROCES STATPL (Y023) is on. However, during PROCES STAGE2 (R040), the R624 timing relay produces a one-second flashing effect on the pilot light. This feature is especially useful in an automated process in which it may appear that nothing is happening, yet a critical timed soak or etch treatment is in process.

HIGH AND LOW LEVEL CIRCUIT

This section describes a "High and Low Level Circuit" that contains several features, including a safety feature and an automatic restart feature. The objective of this circuit is to control a fill DEVICE (Y021) that keeps an in-process bin or tank filled with product as the product is also being drawn out of the bin or tank. The way to remove the product from the bin or tank is not part of this particular circuit. The circuit makes sure the product does not overflow the tank and restarts the fill DEVICE (Y021) before the tank runs out of product. The circuit also assumes that the product supply system can outpace the product removal system.

The product itself can be liquid, powder, grain, or any particulate material. The high-level and the low-level sensors must, of course, be compatible with the material, and have contact outputs.

The circuit contains four inputs: a START push button, a STOP push button, a high-level sensor, and a low-level sensor. It also contains five outputs: a fill DEVICE and four pilot lights. The fill DEVICE can be a pump or a solenoid valve for liquids, or it can be a rotary dump valve, a bucket elevator, or a conveyor for particulate products.

I have included a feature in this circuit that was discussed in the section entitled "Flashing Light Circuits." This feature is a one-second on/off timing sequence flasher (R624) for the low-level pilot light LL PL (Y023). The circuit is designed to work as follows:

1. The low-level pilot light LL PL (Y023) flashes if the product level is below the low-level sensor LO LEV (X001).
2. The system is started by means of the START push button (X000), and the fill DEVICE (Y021) is started unless the high-level sensor HI LEV (X002) is covered with product.
3. The system continues to fill the tank until the high-level sensor HI-LEV (X002) is covered.
4. The timer (T000 with a 5.0-second time value) is included to eliminate spurious signals that can be caused by the temporary sloshing of liquids or the bridging of solid materials over the high-level sensor (X002). The time value of T000 can be modified to fit the operational characteristics of the product.
5. If the high-level sensor (X002) is still covered after five seconds, the fill DEVICE (Y021) is shut off, but the circuit is still live.
6. The fill DEVICE (Y021) remains off until enough product has been removed from the tank to expose the low-level sensor (X001).
7. Once the low-level sensor (X001) is exposed, the fill DEVICE (Y021) is turned on again.

8. This on/off action continues to be controlled by the high-level sensor (X002) and the low-level sensor (X001) until the operation is stopped with the STOP push button (X003).

This system is live for long periods of time when no filling action is taking place. This occurs during every cycle while the product level is being emptied toward the low-level sensor (X001), but neither the FIL PL fill device pilot light (Y022), the HL PL high-level pilot light (Y024), nor the LL PL low-level pilot light (Y023) are on. Therefore, the RUN pilot light (Y020) has been provided as a safety feature to remind the operating personnel that the fill DEVICE (Y021) is in the automatic mode and can restart at any time (Figure 5.15).

Figure 5.15 High and Low Level Circuit

```
|START   STOP                                                   RUN
|-| |--+-|/|----------------------------------------+--------( )-
|X000   |X003                                       |         R000
|       |                                           |
|RUN    |                                           |         RUN PL
|-| |--+                                            +--------( )-
|R000                                                         Y020
|
|RUN     HLSTOP                                               DEVICE
|-| |----|/|----------------------------------------+--------( )-
|R000    R001                                       |         Y021
|                                                   |
|                                                   |         FIL PL
|                                                   +--------( )-
|                                                             Y022
|
|LO LEV FLASHR                                                LL PL
|-|/|----|/|-------------------------------------------------( )-
|X001    R624                                                 Y023
|
|HI LEV                                                       HL PL
|-| |--------------------------------------------------------( )-
|X002                                                         Y024
|
|RUN     HI LEV LO LEV                                        HLSTOP
|-| |--+-| |--+-| |---[00050  TON    T000 ]------------------( )-
|R000   |X002  |X001                                          R001
|       |      |
|       |HLSTOP|
|       +-| |--+
|        R001
|
| +----+
|-| END|
| +----+
```

LEAD-LAG CIRCUITS

This section describes two basic variations of lead-lag circuits. Because of the great diversity of operations in which lead-lag circuits are used, these circuits are included to be shown only as starting-point examples. However, they can serve as circuits you can modify later for use in your own plant as the operational description becomes known.

What is a lead-lag circuit? In a two-output lead-lag circuit, the outputs are switched alternately so that the leading output changes with every other sequence. Let's use outputs Y021 and Y022 that control two pumps as an example. The first time the outputs are energized, Y021 is the lead output and comes on first. Y022 is the lag output and comes on whenever the second pump is required. When the operation is complete, both outputs shut off. The next time the operation starts, the outputs will switch. Y022 becomes the lead output and comes on first. Y021 becomes the lag output and comes on second.

Both examples in this section control three outputs, so they are really lead-lag-last circuits. Let's use outputs Y021, Y022, and Y023 as examples. The first time the outputs are energized, Y021 is the lead output and comes on first, Y022 is the lag output and comes on second, and Y023 is the last output and comes on third. When the operation is complete, all of the outputs shut off. For the next sequence, the outputs will switch. Y022 becomes the lead output and comes on first, Y023 becomes the lag output and comes on second, and Y021 becomes the last output and comes on third. When this operation is complete, all of the outputs shut off. For the third and final sequence, the outputs will switch again. Y023 becomes the lead output and comes on first, Y021 becomes the lag output and comes on second, and Y022 becomes the last output and comes on third. In other words, the changing sequence of outputs is 21-22-23, 22-23-21, and 23-21-22. The fourth sequence of outputs returns to 21-22-23, and the sequences provide a repeating pattern.

The first of the two lead-lag-last circuits presented in this section uses two shift registers. The second circuit is based on relay ladder logic (RLL).

Lead-Lag-Last Circuit

Shift Register Version

To review the way shift registers are loaded into a circuit, or if you need to refresh yourself on how they operate, refer to the section entitled "Shift Registers" in chapter 3.

To test the first circuit, you need just one push button, connected to input X000, which is used as the SHIFT input. This shift register circuit operates three outputs with the one SHIFT push button and in a lead-lag-last circuit mode of operation. Press the X000 SHIFT input push button three times to bring on the first sequence of Y021, then Y022, and then Y023. Energizing the third output in any sequence energizes timer T007, which shuts off the outputs after three seconds. Pressing the X000 SHIFT push button three more times brings on the second sequence of Y022, Y023, and Y021. The third time the X000 SHIFT push button is pressed three times, the sequence is Y023, Y021, and Y022 (Figures 5.16 and 5.17).

Figure 5.16 Lead-Lag-Last Circuit—Shift Register Version (Page 1)

```
|DATA 1                                          +----+
|-| |--------------------------------------------D SR Q
|R010                                            |    |
|                                                |    |
|SHIFT                                           |    |
|-| |----|^|----|/|----|/|----|/|---S(04)|
|X000    R100    Y021    Y022    Y023            |    |
|                                                |    |
|                                                |    |
|------------------------------------------------E R020
|                                                +----+
|
|                                                                  DATA 1
|-|/|----|/|---------------------------------------------------------( )-
|R020    R021                                                      R010
|
|DATA 2                  +----+
|-| |--------------------D SR Q
|R011                    |    |
|                        |    |
|SHIFT                   |    |
|-| |----|^|---S(04)|
|X000    R200            |    |
|                        |    |
|                        |    |
|-|/|--------------------E R030
|R007                    +----+
|
|                                                                  DATA 2
|-|/|----|/|----|/|--------------------------------------------------( )-
|R030    R031    R032                                              R011
|
|-| |----| |--+-|/|--------------------------------------------------( )-
|R020    R030 |R007                                                Y021
|             |
|-| |----| |--|
|R021    R032 |
|             |
|-| |----| |--|
|R022    R031 |
|             |
|-| |---------+
|Y021
```

Figure 5.17 Lead-Lag-Last Circuit—Shift Register Version (Page 2)

```
|-| |----| |--+-|/|-----------------------------------------( )-
|R020   R031  |R007                                        Y022
|             |
|-| |----| |--|
|R021   R030  |
|             |
|-| |----| |--|
|R022   R032  |
|             |
|-| |---------+
|Y022
|
|-| |----| |--+-|/|-----------------------------------------( )-
|R020   R032  |R007                                        Y023
|             |
|-| |----| |--|
|R021   R031  |
|             |
|-| |----| |--|
|R022   R030  |
|             |
|-| |---------+
|Y023
|
|-| |----| |----| |---[00030  TON   T000 ]------------------( )-
|Y021   Y022   Y023                                        R007
|
| +----+
|-| END|
| +----+
```

Relay Ladder Logic Version

The second circuit that provides a three-output lead-lag-last operation is designed with RLL only; no shift registers are used.

To test or experiment with this circuit, you need three push buttons and a two-position selector switch. Connect the selector switch to input X007; it serves as the ON-OFF switch. Connect the three push buttons to inputs X000, X001, and X002. These push buttons simulate three different level sensors in a wet well. The operation of this circuit is the same as that of the shift register version, except the single push button has been replaced by three separate push buttons (Figures 5.18, 5.19, and 5.20).

Figure 5.18 Lead-Lag-Last Circuit—Relay Ladder Logic Version (Page 1)

```
|LEVEL1 RESET                                                   LEAD
|-| |--+-|/|----------------------------------------------------( )-
|X000   |R009                                                   R000
|       |
|LEAD   |
|-| |--+
|R000
|
|LEVEL2 RESET                                                   LAG
|-| |--+-|/|----------------------------------------------------( )-
|X001   |R009                                                   R001
|       |
|LAG    |
|-| |--+
|R001
|
|LEVEL3 RESET                                                   LAST
|-| |--+-|/|----------------------------------------------------( )-
|X002   |R009                                                   R002
|       |
|LAST   |
|-| |--+
|R002
|
|ON-OFF EOS123                                                  SEQ123
|-| |----|/|----------------------------------------------------( )-
|X007   R006                                                    R003
|
|EOS123 EOS231                                                  SEQ231
|-| |----|/|----------------------------------------------------( )-
|R006   R007                                                    R004
|
|EOS231 EOS312                                                  SEQ312
|-| |----|/|----------------------------------------------------( )-
|R007   R008                                                    R005
```

Figure 5.19 Lead-Lag-Last Circuit—Relay Ladder Logic Version (Page 2)

```
|RESET  EOS312                                                 EOS123
|-| |--+-|/|-------------------------------------------------------( )-
|R009   |R008                                                  R006
|       |
|EOS123|
|-| |--+
|R006
|
|SEQ231 RESET  EOS312                                          EOS231
|-| |----| |--+-|/|-----------------------------------------------( )-
|R004   R009  |R008                                            R007
|             |
|EOS231       |
|-| |---------+
|R007
|
|SEQ312 RESET                                                  EOS312
|-| |----| |-------------------------------------------------------( )-
|R005   R009                                                   R008
|
|LAST                                                          RESET
|-| |---[00030   TON    T000 ] ------------------------------------( )-
|R002                                                          R009
|
|SEQ123 LEAD                                                   PUMP 1
|-| |----| |--+----------------------------------------------------( )-
|R003   R000  |                                                Y021
|             |
|SEQ231 LAST  |
|-| |----| |--+
|R004   R002  |
|             |
|SEQ312 LAG   |
|-| |----| |--+
|R005   R001
```

Figure 5.20 Lead-Lag-Last Circuit—Relay Ladder Logic Version (Page 3)

```
|SEQ123 LAG                                                PUMP 2
|-| |----| |--+----------------------------------------------( )-
|R003   R001  |                                            Y022
|             |
|SEQ231 LEAD  |
|-| |----| |--|
|R004   R000  |
|             |
|SEQ312 LAST  |
|-| |----| |--+
|R005   R002
|
|SEQ123 LAST                                               PUMP 3
|-| |----| |--+----------------------------------------------( )-
|R003   R002  |                                            Y023
|             |
|SEQ231 LAG   |
|-| |----| |--|
|R004   R001  |
|             |
|SEQ312 LEAD  |
|-| |----| |--+
|R005   R000
|
| +----+
|-| END|
| +----+
```

6 SPECIAL CIRCUITS

REGISTERS

Registers are probably the single most difficult subject to grasp in the world of programmable logic controllers (PLCs). They definitely produce the most frustration. In this section, we will try to present the subject of registers in such a way as to eliminate a good portion of that frustration.

All of the information within a PLC is stored in registers. That includes the on/off status of each input and each output, the preset and current values of timers and counters, the digital values of the analog inputs and outputs, and all the data for the various advanced function blocks of the PLC. Let's start by defining three terms.

Register the place where information is stored in the microprocessor or central processing unit (CPU) of the PLC.

Bit acronym for *b*inary dig*it*. A bit can be either a zero or a one.

Word the number of bits that a microprocessor or CPU of a PLC can manipulate at one time. Microprocessors typically contain "words" with a standard number of bits, such as eight-bits, sixteen-bits, or thirty-two-bits. The registers in each Toshiba Medium PLC use sixteen-bit words. These sixteen-bit words correspond to the hexadecimal addressing system of 0 through F.

I am convinced that the concept of specifying a register as a word has led to some of the confusion about registers. Let's try to clear this up. PLC registers (or words) do not contain real words as we are accustomed to them. In fact, they do not even contain letters that could make up words. What each register does contain is sixteen bits of information. Remember, each bit is stored as a zero or a one.

If the information stored in the registers is not in real words, what kind of information is stored there? After working with registers and function blocks (see the section entitled "Function Block Programming"), you will begin to see that the type of information stored in a register is impressive. The following are four specific examples of the types of information that can be stored in a register:

1. Whether an input or an output is on or off.
2. The present value of the analog pressure (in digital form) of a pump as read by a pressure sensor.
3. The quantity of parts produced by your punch press today.
4. The quantity of parts produced by your punch press today, divided by twelve to measure the production rate in dozens of parts.

As can be seen from these four examples, two very different types of information can exist in registers. The first type of information is *status*, and the second type of information is *data*.

And all registers, of course, have addresses. Table 6.1 contains a list of the various PLC devices, their beginning register addresses, and the types of information (status or data) that they can contain.

PLC registers do not contain both types of information at the same time. They can contain either status or data, but not both.

Only four of the register types can contain status information. They are related to relays, inputs, outputs, and link devices when they are used as digital devices. *Status* refers to the state of that device—whether it is on and its bit is a one (1), or whether it is off and its bit is a zero (0).

A register with data information in it contains numbers that represent items such as well levels, pressures, production counts, and timer values. All seven register types can contain data information. When any of the four registers mentioned in the previous paragraph (relays, inputs, out-

Table 6.1 PLC Devices, Beginning Register Addresses, and Types of Information They Contain

Device Description	*Beginning Register Address*	*Type of Information*
Relays (internal)	RW00	Status or data
Inputs (analog and digital)	XW00	Status or data
Outputs (analog and digital)	YW00	Status or data
Link devices	ZW00	Status or data
Counters	C00	Data only
Data	D000	Data only
Timers	T000	Data only

puts, and link devices) is used as an analog device, it contains data information.

Let's turn now to the addresses for the individual digital devices within these register addresses. Each register has one more digit added to it that represents one of sixteen bits. For example, if your START push button is to be hard-wired to input terminal X3 of an M40 PLC, the on/off status of that push button is stored as a one or a zero in the third position in register XW00. Similarly, if your STOP push button is to be hard-wired to input terminal XF of this M40 PLC, the on/off status of that push button is stored as a one or a zero in the sixteenth, or F, position in register XW00.

Therefore, the address of an individual digital device is the combination of the register number (leaving out the W) and the bit position number. This rule applies to all inputs, outputs, relays, and link devices.

Let's look at the last two paragraphs and tie up the loose ends. If we look at the START push button that is hard-wired to input terminal X3, and drop the W from the register number XW00, we arrive at an address for that input of X003. Likewise, if we look at the STOP push button that is hard-wired to input terminal XF, and drop the W from the register number XW00, we arrive at an address for that input of X00F.

The information that is stored as data in registers that apply to analog values, timers, and counters is in the form of a binary zero or one, but can be viewed as either hexadecimal or decimal. You can view the contents of any sixteen-bit register (status or data) by pressing F3–Block Monitor from the TOP MENU.

Registers are discussed further in the next section, entitled "I/O Allocation" and in the section entitled "Function Block Programming." These two sections show the many ways in which information in the various registers can be manipulated. Registers are used extensively in the examples in the rest of this chapter. Additional discussions about registers can be found in the manuals for each of the PLCs.

I/O ALLOCATION

Input-Output (I/O) Allocation deals with the installation of your I/O modules into their racks of the EX100, the EX250, and the EX500 PLCs. It also deals with installation of the I/O modules into the expansion rack for the M20 and the M40 PLCs. It also involves the assignment of the registers and addresses for these I/O modules.

You probably want to know why any of this is important to you. The discussion in this whole section really boils down to one fact: *Your pro-*

gram will not work in your PLC until you have performed all the steps of the operation called I/O Allocation. This may raise additional questions from those of you who have worked with the Toshiba Small PLCs, because those PLCs do *not* require that you perform the I/O Allocation.

What is the difference between the small and the medium series of Toshiba PLCs that causes you to perform this extra step in the programming of the Medium PLCs? The main difference is that the medium series is much more flexible in regard to inputs and outputs than is the small series.

For example, in the small series of PLCs, the inputs and outputs are limited to a relatively small quantity of digital devices. Toshiba has established permanent locations for these devices on the PLC and has assigned their addresses. As a result, each input and output terminal is labeled with an address on the PLC housing.

In contrast to this, the Medium PLCs can handle a large number of analog, digital, and special modules. Because there can be a great variation in the use of these modules from project to project, these modules cannot be fixed in their locations, nor can they be assigned permanent addresses by Toshiba. In fact, the modules are installed in racks and can be installed within the racks in a wide variety of arrangements. Therefore, installation of the modules in the racks and assignment of their addresses is the responsibility of the user. It is accomplished by means of the I/O Allocation.

Although there are no rigid rules for installing I/O modules in the racks, it is advisable to install the analog input modules and the analog output modules next to each other so that the wiring for these modules can be kept together and can be isolated from the higher voltage wires of the other modules. You can even leave an empty slot between modules in your rack. However, you must also leave a gap in the assignment of registers in the I/O Allocation operation to account for the gap.

Now let's get into the subject of I/O Allocation in depth. For the purposes of I/O Allocation, I/O modules are defined by type and size. There are only four types of I/O modules. The first type is for inputs, and is designated with an *X*. The second type is for outputs, and is designated with a *Y*. The third type is for the axis motion control module and the ASCII/BASIC module, and is designated by *X+Y*. The fourth type is for the TOSLINE-30 module, and is designated with a *Z*. Only the X and Y types of modules will be discussed in this book.

There are six sizes of I/O modules. These six sizes are defined as *words* (see the section entitled "Registers"), and are designated as one-word, two-word, four-word, eight-word, sixteen-word, and thirty-two–word registers. These words are also abbreviated as 1W, 2W, 4W, 8W, 16W, and 32W. However, only one-word, two-word, and four-word registers will be discussed in this book.

To add to the confusion, types and sizes can be intermixed. But stay with me, and I'll try to sort out all of this in such a way as to make it understandable. Within the types and sizes discussed in this book, you will find the following mix:

Inputs	X 1W	X 2W	X 4W
Outputs	Y 1W	Y 2W	Y 4W

To ease the confusion, each module has its own type and size designation, which can be found in the instructions packaged with each module.

For those of you who are working with M20, M40, or the EX100 PLCs, a chart in chapter 6 of your PLC manual lists all of the modules that can be used with that particular PLC. The most important part of that chart is the column entitled *Module Type,* which contains the module type and size information. The chart for the M20 and the M40 PLC can be found on page 92 of its manual. The chart for the EX100 PLC can be found on page 76 of its manual.

To see how this I/O Allocation operation works, let's take three examples of an EX100 PLC with two input and two output modules and change the sizes of the modules.

In the first example, all of the modules are sixteen-point, one-word modules. The I/O Allocation is easy; starting with 00, each module receives a register number that increases by one as you go up the line of modules. Each register number is preceded by an *XW* or a *YW.* Table 6.2 summarizes the I/O Allocation by slot number.

In the second example, all four of the sixteen-point, one-word modules are replaced with thirty-two-point, two-word modules. The I/O Allo-

Table 6.2 EX100 I/O Allocation Chart

Slot No.	*1*	*2*	*3*	*4*	*5*	*6*
Module	P/S	CPU	Input	Input	Output	Output
Register No.	–	–	XW00	XW01	YW02	YW03
Address No.	–	–	X000 thru X00F	X010 thru X01F	Y020 thru Y02F	Y030 thru Y03F

cation is still easy, because each module receives a register with two sets of numbers, starting with 00. Table 6.3 summarizes I/O Allocation by slot number.

In the third example, all four of the thirty-two-point, two-word modules are replaced. The following four modules are used:

1. One sixteen-point, one-word input module
2. One four-channel, four-word analog input module
3. One two-channel, two-word analog output module
4. One twelve-point, one-word relay output module

Table 6.4 summarizes the I/O Allocation.

These three examples show how any arrangement of modules in a rack can be set up for the I/O Allocation.

Just remember that there are two basic steps to the I/O Allocation operation. The first step is to install the modules in the rack in any order that you choose. The second step is to use a chart to assign the registers and addresses, as shown in the charts of the examples presented in this chapter. The chart is an aid to guide you in your programming and in assigning addresses to the various I/O devices in your circuit.

Once you have established the I/O Allocation for your project, it must be entered into the program. To enter the I/O Allocation into your program, go to the TOP MENU and press F1–System and Control Information. Press F1–System Information and type your program name on the first line. Move the cursor down to line six, and press the space bar until

Table 6.3 EX100 I/O Allocation Chart

Slot No.	*1*	*2*	*3*	*4*	*5*	*6*
Module	P/S	CPU	Input	Input	Output	Output
Register No.	–	–	XW00	XW02	YW04	YW06
			XW01	XW03	YW05	YW07
Address No.	–	–	X000	X020	Y040	Y060
			thru	thru	thru	thru
			X00F	X02F	Y04F	Y06F
			and	and	and	and
			X010	X030	Y050	Y070
			thru	thru	thru	thru
			X01F	X03F	Y05F	Y07F

Table 6.4 EX100 I/O Allocation Chart

Slot No.	*1*	*2*	*3*	*4*	*5*	*6*
Module	P/S	CPU	Input	Input (Analog)	Output (Analog)	Output
Register No.	–	–	XW00	XW01 XW02 XW03 XW04	YW05 YW06	YW07
Address No.	–	–	X000 thru X00F	XW01 XW02 XW03 XW04	YW05 YW06	Y070 thru Y07B

your PLC type is displayed. Press F10 to save this information and return to the previous screen.

Press F2–I/O Allocation. The new screen allows you to specify which type of module you have installed in each slot, based on the chart that you prepared.

Move the cursor down to the first open slot. From the command line, select the type of module that you installed in that slot, and press the F key that corresponds to it. For example, if your first module is an input module, press F1–Input X. The slot is then filled with X 1W. Should your module require an X 2W or an X 4W, press the space bar until that word size appears. Use the cursor and the space bar to repeat the process for all of the other modules. When you are finished, press F10 twice to save this information and to return to the TOP MENU.

For more information about I/O Allocation, refer to chapter 6 of your PLC manual.

FUNCTION BLOCK PROGRAMMING

Function block programming is an extension of relay ladder logic (RLL) that allows you to perform many complex functions. Thus it adds a great deal to the capabilities of your RLL circuits.

Because of the variety and number of function blocks available, the chapter entitled "Instruction" of your PLC manual should be helpful to you. That "Instruction" chapter provides two sections about function blocks, and these two sections are tied together by the FUN. NO. for each function block. The first section of chapter 7 in the PLC manual gives the function block its FUN. NO. and includes a descriptive name, as well as a brief description of what each function block does. The second section provides a full page of instructions on the operation of each function block and how the registers operate to obtain the desired results.

Sixty-seven different function blocks are available in the Toshiba Medium PLCs. They have been assigned to the six groups in Table 6.5 (the column labeled *Command Line 2* is discussed later).

Read the brief descriptions of these function blocks in the first section of the "Instruction" chapter in your PLC manual. This gives you an idea of the magnitude and variety of the operations that can be performed within your RLL circuits.

When you need more information about a particular function block, consult the second section of the "Instruction" chapter in your PLC manual. There the function blocks are arranged by FUN. NO., and each function block is provided with a full page (or more) of instructions on its operation, along with a description on how the registers operate to obtain the desired results.

When you are ready to load a function block into the circuit, refer to the *Command Line 2* column in Table 6.5.

Table 6.5 Function Blocks Available in the Toshiba Medium PLCs

Function Type Group	*FUN. NO.*	*Command Line 2*
Data transfer instructions	FUN.000 through FUN.005	F3–Move Comp
Arithmetic operations	FUN.010 through FUN.013	F4–Arith
	FUN.014 through FUN.016	F3–Move Comp
	FUN.017 through FUN.018	F4–Arith
	FUN.020 through FUN.023	F4–Arith
	FUN.024 through FUN.026	F3–Move Comp
Logical operations	FUN.030 through FUN.032	F5–Logic Oper
	FUN.034 through FUN.036	F5–Logic Oper
	FUN.040 through FUN.043	F5–Logic Oper
	FUN.046	F5–Logic Oper
Data conversion instructions	FUN.050 through FUN.055	F6–Conv. Limit
Special functions	FUN.060 through FUN.065	F6–Conv. Limit
	FUN.070 through FUN.074	F4–Arith
Other functions	FUN.080 through FUN.081	F2–Ext. Seq.
	FUN.090 through FUN.091	F3–Move Comp
	FUN.096 through FUN.097	F6–Conv. Limit
	FUN.100 through FUN.102	F2–Ext. Seq.
	FUN.110 through FUN.112	F2–Ext. Seq.

Figure 6.1 Command Line 2

```
 Seq. Ext.  Move          Logic  Conv.  Insert  Move  Assign   1Acc.
1Ins 2Seq. 3Comp 4Arith 5Oper 6Limit 7Delete 8Copy 9Synonym 0
```

Note: The column labeled *Command Line 2* is included in the table because the EX-PDD software program does not use the FUN. NO. to specify the function blocks. The EX-PDD software puts all of the function blocks into the five groups of F2 through F6 in command line 2.

Command line 2 is shown in Figure 6.1.

Let's take FUN. NO. 25 as an example, and see what steps must be followed to load this function block into your circuit. The name of FUN. NO. 25 is Constant Comparison Equal. The purpose of this register is to compare the number in a register with a number that you can select and that is a constant number. The number in the register is changing during the operation of your system. When the number in the register equals the constant, the output of the function block is turned on. This comparison of data in the register to the constant number takes place with every scan of the program.

You are now at the starting point for loading a function block into a circuit. While in the Edit Page mode, go to command line 2. Table 6.5 shows that you can get to FUN. NO. 25 by pressing F3–Move Comp while at command line 2. You are given a new command line that presents a number of function block options. When you press Shift-F7–R=K, the device shown in Figure 6.2 is placed on your screen at the point of the cursor.

This is the block symbol for FUN. NO. 25, the Constant Comparison Equal function block. The cursor appears in the left-hand portion of the function block. Type in the address of the register (**D0140** in this example) where the data can be found. Press ENTER and move the cursor over to the right-hand portion of the function block. Type in the number for the constant (**335** in this example), and press ENTER again. The function block will look like Figure 6.3.

Figure 6.2 Block Symbol for FUN. NO. 25, Constant Comparison Equal

```
|
|-[        = .        ]
|
```

Figure 6.3 Function Block after Address Register and Constant Have Been Entered

```
|
|-[D0140   = .   00335]
|
```

Figure 6.4 Function Block Circuit after ALARM Output Has Been Entered

```
|                                                                     ALARM
|-[D0140   = .   00335]-----------------------------------------------( )-
|                                                                     Y020
```

Add ALARM output Y020 so that the circuit looks like Figure 6.4.

In this circuit the ALARM is normally off. Whenever the data in the register D0140 increases to the point where it equals a value of 335, the output turns on, and ALARM Y020 is energized. All function blocks are programmed in a similar way.

In summary, refer to the "Instruction" chapter (chapter 7) in your PLC manual. Select the FUN. NO. of the function block you need from the first section of the chapter, and then find that FUN. NO. in the second section for a complete explanation of the function block. Refer to Table 6.5 in this section of the *Toshiba Medium PLC Primer* for the F key that installs that function block onto your personal computer (PC) screen.

Function blocks present an unusual situation for RLL programmers. They do not require a coil-type output. Such a circuit looks strange to someone accustomed to having a complete rung spread across the page to the right-hand power line. This feature of function blocks will become more apparent in the sections that follow, in which specific applications of functions blocks are described.

USING THE CLOCK-CALENDAR IN A CIRCUIT

The M20, the M40, and the EX100 PLCs that contain the enhanced version of the CPU can be programmed to use their clock-calendars to keep track of the year, month, day, day of the week, hour, minute, and second, with an accuracy of plus or minus thirty seconds per month. However, only a few EX250 and EX500 PLCs were sold that contained the enhanced version of the CPU.

The circuit presented in this section is a very simple circuit, and has been included only to whet your appetite. This circuit compares data

Figure 6.5 Rung Used to Reserve Registers for Clock-Calendar Data

```
|
|-| |------------------------------------------------( )-
|R62F                                             R628
```

with the clock-calendar data registers and sounds the lunch whistle every day at 11:59 A.M. for fifteen seconds. This simple operation contains enough information to show you how to set up and use the clock-calendar in a circuit to obtain a very "timely" operation.

The circuit uses the Constant Comparison Equal function block (FUN. NO. 25) to compare the actual time stored in the clock-calendar registers with the hour and minutes that you specify for sounding the lunch whistle. To use this feature in any circuit, data registers D0005 through D0010 must be reserved for the clock-calendar data. That is done automatically by inserting the rung in Figure 6.5 on the first page of your circuit.

The six clock-calendar functions and their corresponding data registers are reserved as follows:

Function	*Register*
Year	D0005
Month	D0006
Week/Day	D0007
Hour	D0008
Minute	D0009
Second	D0010

All of the data for these data registers must be in two-digit binary coded decimal (BCD) code. For example, the BCD codes for the actual time are based on the twenty-four hour clock and are broken down into register D0008 for the hours and register D0009 for the minutes. For the BCD codes for the year, the month, and the day, refer to Table 6.6.

To make this a little more understandable, let's go through the steps that are required to enter the clock-calendar data into the appropriate data registers. Imagine that today's date is January 15, 1996, and that the time is 11:59 A.M. To perform this operation, use the following steps:

1. Start at the TOP MENU. Press F7–Enter/Change program name. Overwrite the word *default* with a new program name, and press F10 to save that name and to return to the TOP MENU.

Table 6.6 BCD Codes for the Clock-Calendar

Year	*BCD Code*	*Month*	*BCD Code*	*Day*	*BCD Code*
1995	95	Jan.	01	Sun.	00
1996	96	Feb.	02	Mon.	01
1997	97	Mar.	03	Tue.	02
1998	98	Apr.	04	Wed.	03
1999	99	May	05	Thu.	04
		June	06	Fri.	05
		July	07	Sat.	06
		Aug.	08		
		Sep.	09		
		Oct.	10		
		Nov.	11		
		Dec.	12		

2. Press F3–Block Monitor, which takes you to a block monitor screen that looks like Table 6.7.
3. Press F1–Start Number, and a message appears that reads as follows:

   `Enter the new starting register number: YW00.`

4. Overwrite the YW00 with **D0005**, and press ENTER.
5. The screen changes, and register D0005 is at the top of the Register (Reg.) column. The screen looks like Table 6.8.
6. Press F2–DEC, and it will change to read F2–HEX. At the same time, all the numbers in the Value column change to H0000, and the screen looks like Table 6.9.

   Note: Clock-calendar numbers entered into the data registers must be entered in the Hex mode of the block monitor screen. At the block monitor screen, press F2 to change from DEC to HEX. For example, to enter the *12* for *December* into the D0006 data register, type **H0012**.

7. Press F5–Set value, and a message appears that reads as follows:

   `Register D0005 ... Enter New Value: 00000.`

8. Register D0005 is the register for the year, so overwrite the 00000 with **H0096** (for 1996), and press ENTER.
9. A confirmation message appears near the bottom of the screen, and the number H0096 appears in the Value column. The value for 96 also appears in the appropriate register bit columns of F through 0.

Table 6.7 Block Monitor at Beginning of Operation

Reg.	Value	FEDC	BA98	7654	3210	Reg.	Value	FEDC	BA98	7654	3210
YW00	00000	0000	0000	0000	0000	YW16	00000	0000	0000	0000	0000
YW01	00000	0000	0000	0000	0000	YW17	00000	0000	0000	0000	0000
YW02	00000	0000	0000	0000	0000	YW18	00000	0000	0000	0000	0000
YW03	00000	0000	0000	0000	0000	YW19	00000	0000	0000	0000	0000
YW04	00000	0000	0000	0000	0000	YW20	00000	0000	0000	0000	0000
YW05	00000	0000	0000	0000	0000	YW21	00000	0000	0000	0000	0000
YW06	00000	0000	0000	0000	0000	YW22	00000	0000	0000	0000	0000
YW07	00000	0000	0000	0000	0000	YW23	00000	0000	0000	0000	0000
YW08	00000	0000	0000	0000	0000	YW24	00000	0000	0000	0000	0000
YW09	00000	0000	0000	0000	0000	YW25	00000	0000	0000	0000	0000
YW10	00000	0000	0000	0000	0000	YW26	00000	0000	0000	0000	0000
YW11	00000	0000	0000	0000	0000	YW27	00000	0000	0000	0000	0000
YW12	00000	0000	0000	0000	0000	YW28	00000	0000	0000	0000	0000
YW13	00000	0000	0000	0000	0000	YW29	00000	0000	0000	0000	0000
YW14	00000	0000	0000	0000	0000	YW30	00000	0000	0000	0000	0000
YW15	00000	0000	0000	0000	0000	YW31	00000	0000	0000	0000	0000

Table 6.8 Block Monitor after Entry of Starting Register Number

Reg.	Value	FEDC	BA98	7654	3210	Reg.	Value	FEDC	BA98	7654	3210
D0005	00000	0000	0000	0000	0000	D0021	00000	0000	0000	0000	0000
D0006	00000	0000	0000	0000	0000	D0022	00000	0000	0000	0000	0000
D0007	00000	0000	0000	0000	0000	D0023	00000	0000	0000	0000	0000
D0008	00000	0000	0000	0000	0000	D0024	00000	0000	0000	0000	0000
D0009	00000	0000	0000	0000	0000	D0025	00000	0000	0000	0000	0000
D0010	00000	0000	0000	0000	0000	D0026	00000	0000	0000	0000	0000
D0011	00000	0000	0000	0000	0000	D0027	00000	0000	0000	0000	0000
D0012	00000	0000	0000	0000	0000	D0028	00000	0000	0000	0000	0000
D0013	00000	0000	0000	0000	0000	D0029	00000	0000	0000	0000	0000
D0014	00000	0000	0000	0000	0000	D0030	00000	0000	0000	0000	0000
D0015	00000	0000	0000	0000	0000	D0031	00000	0000	0000	0000	0000
D0016	00000	0000	0000	0000	0000	D0032	00000	0000	0000	0000	0000
D0017	00000	0000	0000	0000	0000	D0033	00000	0000	0000	0000	0000
D0018	00000	0000	0000	0000	0000	D0034	00000	0000	0000	0000	0000
D0019	00000	0000	0000	0000	0000	D0035	00000	0000	0000	0000	0000
D0020	00000	0000	0000	0000	0000	D0036	00000	0000	0000	0000	0000

A confirmation message appears after you have entered each number into its data register.

10. Press F4–Inc & Set, and a message appears that reads as follows:

    `Register D0006 ... Enter New Value: 00000.`

11. Register D0006 is the register for the month, so overwrite the 00000 with **H0001** (for January), and press ENTER.
12. Press F4–Inc & Set again, and a message appears that reads as follows:

    `Register D0007 ... Enter New Value: 00000.`

13. Register D0007 is the register for the day, so overwrite the 00000 with **H0015** (for the 15th), and press ENTER.
14. Press F4–Inc & Set again, and a message appears that reads as follows:

    `Register D0008 ... Enter New Value: 00000.`

15. Register D0008 is the register for the hour, so overwrite the 00000 with **H0011** (for the 11:00 portion of 11:59), and press ENTER.

Table 6.9 Block Monitor after Change to Hexadecimal Mode

Reg.	Value	FEDC	BA98	7654	3210	Reg.	Value	FEDC	BA98	7654	3210
D0005	H0000	0000	0000	0000	0000	D0021	H0000	0000	0000	0000	0000
D0006	H0000	0000	0000	0000	0000	D0022	H0000	0000	0000	0000	0000
D0007	H0000	0000	0000	0000	0000	D0023	H0000	0000	0000	0000	0000
D0008	H0000	0000	0000	0000	0000	D0024	H0000	0000	0000	0000	0000
D0009	H0000	0000	0000	0000	0000	D0025	H0000	0000	0000	0000	0000
D0010	H0000	0000	0000	0000	0000	D0026	H0000	0000	0000	0000	0000
D0011	H0000	0000	0000	0000	0000	D0027	H0000	0000	0000	0000	0000
D0012	H0000	0000	0000	0000	0000	D0028	H0000	0000	0000	0000	0000
D0013	H0000	0000	0000	0000	0000	D0029	H0000	0000	0000	0000	0000
D0014	H0000	0000	0000	0000	0000	D0030	H0000	0000	0000	0000	0000
D0015	H0000	0000	0000	0000	0000	D0031	H0000	0000	0000	0000	0000
D0016	H0000	0000	0000	0000	0000	D0032	H0000	0000	0000	0000	0000
D0017	H0000	0000	0000	0000	0000	D0033	H0000	0000	0000	0000	0000
D0018	H0000	0000	0000	0000	0000	D0034	H0000	0000	0000	0000	0000
D0019	H0000	0000	0000	0000	0000	D0035	H0000	0000	0000	0000	0000
D0020	H0000	0000	0000	0000	0000	D0036	H0000	0000	0000	0000	0000

16. Press F4–Inc & Set one last time, and a message appears that reads as follows:

    `Register D0009 ... Enter New Value: 00000.`

17. Register D0009 is the register for the minutes, so overwrite the 00000 with **H0059** (for the 00:59 portion of 11:59), and press ENTER.
18. Don't bother to enter data into the D0010 register for seconds. This cannot be accomplished with any degree of accuracy until after you have gone on-line.
19. When all of the data has been loaded, the block monitor screen will look like Table 6.10.
20. To save this data into the registers and to return to the TOP MENU, press F10.

The circuit is shown in Figure 6.6.

Two final notes apply to the clock-calendar. First, when the power to the PLC is turned off, the clock-calendar is updated by a built-in capaci-

Table 6.10 Block Monitor after All Data Has Been Loaded

Reg.	Value	FEDC	BA98	7654	3210	Reg.	Value	FEDC	BA98	7654	3210
D0005	H0096	0000	0000	1001	0110	D0021	H0000	0000	0000	0000	0000
D0006	H0001	0000	0000	0000	0001	D0022	H0000	0000	0000	0000	0000
D0007	H0015	0000	0000	0001	0101	D0023	H0000	0000	0000	0000	0000
D0008	H0011	0000	0000	0001	0001	D0024	H0000	0000	0000	0000	0000
D0009	H0059	0000	0000	0101	1001	D0025	H0000	0000	0000	0000	0000
D0010	H0000	0000	0000	0000	0000	D0026	H0000	0000	0000	0000	0000
D0011	H0000	0000	0000	0000	0000	D0027	H0000	0000	0000	0000	0000
D0012	H0000	0000	0000	0000	0000	D0028	H0000	0000	0000	0000	0000
D0013	H0000	0000	0000	0000	0000	D0029	H0000	0000	0000	0000	0000
D0014	H0000	0000	0000	0000	0000	D0030	H0000	0000	0000	0000	0000
D0015	H0000	0000	0000	0000	0000	D0031	H0000	0000	0000	0000	0000
D0016	H0000	0000	0000	0000	0000	D0032	H0000	0000	0000	0000	0000
D0017	H0000	0000	0000	0000	0000	D0033	H0000	0000	0000	0000	0000
D0018	H0000	0000	0000	0000	0000	D0034	H0000	0000	0000	0000	0000
D0019	H0000	0000	0000	0000	0000	D0035	H0000	0000	0000	0000	0000
D0020	H0000	0000	0000	0000	0000	D0036	H0000	0000	0000	0000	0000

tor, which can back up the clock-calendar for seven days. If the power is not turned back on before this backup period is over, the data in the clock-calendar registers may be lost. If the data is lost, it must be re-entered. A backup battery option is available from Toshiba that backs up the clock-calendar for two years.

Second, any date and time that you enter into the block monitor while you are programming the circuit may not be correct when you load the program into the PLC. This can be true even if the last thing that you did in the program was to load the date and time data. The reason for this is that the steps required to load the circuit from the PC to the PLC can take enough time to make the time incorrect. In addition to that, hours or even days may pass between the time that you load the data and the time that you take the circuit on-line.

Once you are on-line and have loaded the circuit into the PLC, stay in the HALT mode and go to the block monitor screen once again. Compare the data you originally entered with the current date and time and make the necessary adjustments—in the HEX mode, of course.

Figure 6.6 Lunch-Hour Whistle Circuit

```
|-| |-------------------------------------------------------( )-
|R62F                                                        R628
|
| HOUR                  MINUTE                               LUNCH
|-[D0008   = .  H0011]-[D0009   = .  H0059]-----------------( )-
|                                                            R000
|
|LUNCH                                                      15 SEC
|-| |--+[00150  TON   T000 ]---------------------------------( )-
|R000  |                                                     R001
|      |
|      |15 SEC                                              WHISTL
|      +-|/|------------------------------------------------( )-
|       R001                                                 Y020
|
| +----+
|-| END|
| +----+
```

STEP SEQUENCER

This section presents a circuit called the "Step Sequencer," which performs the same operations as an electromechanical (motor-driven) drum sequencer. The "Step Sequencer" circuit has seven steps (0–6), and it requires the use of seven inputs to control all of the different outputs in the stepped sequence.

To test this circuit, connect a two-position selector switch to input X000. Connect six push buttons to inputs X001 through X006.

The circuit also uses the following three step sequence function blocks:

FUN. NO. 100	Step Sequence Initialize
FUN. NO. 101	Step Sequence Input
FUN. NO. 102	Step Sequence Output

This circuit operates in the following way.

When ON/OFF selector switch X000 is turned on, the Step Sequence Initialize function block is turned on. Turning on FUN. NO. 100 sets up the step sequencer for the seven steps of R500 through R506 that we have selected for this circuit. Notice that when EX-PDD loads the step sequence function blocks into a rung of the circuit, the addresses for these function blocks are put into brackets like this: [R505].

Turning on the Step Sequence Initialize function block turns on Step Sequence Input R500, which does not turn on any outputs but sets up the circuit for Step 1 [R501]. Pressing the X001 input push button operates Step Sequence Output function block R501, which turns on output Y020.

Turning on Step Sequence Output R501 also turns on Step Sequence Input R501. Step Sequence Input R501 does not turn on any outputs, but sets up the circuit for Step 2 [R502]. Pressing the X002 input push button operates Step Sequence Output function block R502, which turns on outputs Y021, Y023, and Y025.

Turning on Step Sequence Output R502 also turns on Step Sequence Input R502. Step Sequence Input R502 does not turn on any outputs but sets up the circuit for Step 3 [R503]. Pressing and holding down the X003 push button until timer T000 times out at ten seconds operates Step Sequence Output function block R503, which turns on outputs Y020, Y022, and Y024.

Turning on Step Sequence Output R503 also turns on Step Sequence Input R503. Step Sequence Input R503 does not turn on any outputs, but sets up the circuit for Step 4 [R504]. Pressing the X004 input push button operates Step Sequence Output function block R504, which turns on outputs Y020 and Y025.

Turning on Step Sequence Output R504 also turns on Step Sequence Input R504. Step Sequence Input R504 does not turn on any outputs but sets up the circuit for Step 5 [R505]. In this step, however, the X005 input push button must be pressed five times (CNT 00) before it operates Step Sequence Output function block R505, which turns on outputs Y021, Y022, Y024, and Y025.

Turning on Step Sequence Output R505 also turns on Step Sequence Input R505. Step Sequence Input R505 does not turn on any outputs, but sets up the circuit for Step 6 [R506]. Pressing the X006 input push button operates Step Sequence Output function block R506, which turns on all of the outputs in a flashing mode.

Turning on Step Sequence Output R506 also turns on Step Sequence Input R506. Step Sequence Input R506 does not turn on any outputs, but it resets the circuit back to Step 0 after the ten seconds set in timer T001. This shuts off all of the outputs and prepares the circuit for another cycle.

At the end of the sequence, Step Six turns on all the outputs and puts them into a flashing mode. To do this, a normally open (NO) contact, R622, has been inserted in each of the rungs where Step Six (R506) controls the six outputs. The R622 timer is part of a special series of timing devices, and it causes each output in the circuit to flash on and off at a continuous rate of four-tenths of a second per cycle. For more information on these special timing devices, refer to the section entitled "Flashing Light Circuits." Table 6.11 summarizes the circuit operation.

The circuit is shown in Figures 6.7, 6.8, and 6.9.

This circuit can be modified to allow you to test it automatically without requiring the hard-wiring of the six input switches X001 through X006. To do this, the two-position selector switch BYP.SS has been added and has been connected to input X00F. An NO contact from switch BYP.SS has been added as a bypass around each of the step inputs, along with a timer set for five seconds.

To start this modified circuit, turn on both the ON-OFF X000 and the BYP.SS X00F selector switches. The program bypasses all input switches, the two timers, and the counter and automatically steps through the complete sequence in five-second increments. It continues to repeat the cycle until the circuit is shut off.

Table 6.11 Step Sequencer Operational Chart

Inputs	*Step No.*	*Step Output*	*Outputs* *Y020*	*Y021*	*Y022*	*Y023*	*Y024*	*Y025*
X0 Start	0	R500	Off	Off	Off	Off	Off	Off
X1 Advance	1	R501	On	Off	Off	Off	Off	Off
X2 Advance	2	R502	Off	On	Off	On	Off	On
X3 Delay 10 seconds then advance	3	R503	On	Off	On	Off	On	Off
X4 Advance	4	R504	On	Off	Off	Off	Off	On
X5 Advance after 5 counts of X5	5	R505	Off	On	On	Off	On	On
X6 Advance	6	R506	All outputs flash on and off Return to Step 0 after 10 seconds.					

Figure 6.7 Step Sequencer Circuit (Page 1)

```
|ON-OFF
|-| |----|^|--+[STIZ  [07]   R500 ]
|X000   R300  |
|             |
|             |
|-| |----|^|--+
|R000   R301
|
|
|-[R500]--| |------------------------------------------[R501]
|        X001
|
|
|-[R501]--| |------------------------------------------[R502]
|        X002
|
|
|-[R502]--| |---[00100  TON   T000 ]-------------------[R503]
|        X003
|
|
|-[R503]--| |------------------------------------------[R504]
|        X004
|
|
|-[R504]--| |------------------------------------------[R505]
|        R001
|
|
|-[R505]--| |------------------------------------------[R506]
|        X006
```

Figure 6.8 Step Sequencer Circuit (Page 2)

```
|
|-[R506]-[00100   TON    T001 ]---------------------------------( )-
|                                                               R000
|
|         +-----------+
|-| |---C   CNT       Q-----------------------------------------( )-
|X005     |           |                                         R001
|         |           |
|         |           |
|-|/|---E 00005 C00   |
|R000     +-----------+
|
|-| |---------+-------------------------------------------------( )-
|R501         |                                                 Y020
|             |
|-| |---------+
|R503         |
|             |
|-| |---------+
|R504         |
|             |
|-| |----| |--+
|R506    R622
|
|-| |---------+-------------------------------------------------( )-
|R502         |                                                 Y021
|             |
|-| |---------+
|R505         |
|             |
|-| |----| |--+
|R506    R622
|
|-| |---------+-------------------------------------------------( )-
|R503         |                                                 Y022
|             |
|-| |---------+
|R505         |
|             |
|-| |----| |--+
|R506    R622
```

To shut off the circuit completely, turn off both the X000 and the X00F selector switches.

To stop the circuit and hold it at a certain step, turn off BYP.SS selector switch X00F when the sequence reaches that particular step.

The modified circuit is shown in Figures 6.10, 6.11, and 6.12.

Figure 6.9 Step Sequencer Circuit (Page 3)

```
|
|-| |---------+-------------------------------------------( )-
|R502         |                                          Y023
|             |
|             |
|-| |----| |--+
|R506   R622
|
|
|-| |---------+-------------------------------------------( )-
|R503         |                                          Y024
|             |
|             |
|-| |---------+
|R505         |
|             |
|             |
|-| |----| |--+
|R506   R622
|
|
|-| |---------+-------------------------------------------( )-
|R502         |                                          Y025
|             |
|             |
|-| |---------+
|R504         |
|             |
|             |
|-| |---------+
|R505         |
|             |
|             |
|-| |----| |--+
|R506   R622
|
|  +----+
|-| END|
|  +----+
```

BCD THUMBWHEEL CIRCUIT

This section presents two circuits that use a BCD thumbwheel switch in conjunction with function blocks. Both of these circuits have been set up with an EX100 PLC in mind, but they can be used with any Medium PLC if you install the modules in the slots that fit the addresses assigned in the circuit.

Thumbwheel switches come in a wide variety of types, such as BCD, decimal, hexadecimal, octal, and many others. Only the BCD thumbwheel switch type is discussed in this book.

Each section of a thumbwheel switch can be rotated through the standard ten digits of zero through nine. However, each section of a BCD thumbwheel switch has only five leads. One serves as the common con-

Figure 6.10 Step Sequencer Circuit—Modified (Page 1)

```
|ON-OFF
|-| |----|^|--+[STIZ  [07]   R500 ]
|X000   R300  |
|             |
|             |
|-| |----|^|--+
|R000   R301
|
|-[R500]+-| |---------------------+---------------------[R501]
|        |X001                    |
|        |                        |
|        |BYP.SS                  |
|        +-| |---[00050  TON  T101 ]+
|         X00F
|
|-[R501]+-| |---------------------+---------------------[R502]
|        |X002                    |
|        |                        |
|        |BYP.SS                  |
|        +-| |---[00050  TON  T102 ]+
|         X00F
|
|-[R502]+-| |---[00100  TON  T000 ]+---------------------[R503]
|        |X003                    |
|        |                        |
|        |BYP.SS                  |
|        +-| |---[00050  TON  T103 ]+
|         X00F
|
|-[R503]+-| |---------------------+---------------------[R504]
|        |X004                    |
|        |                        |
|        |BYP.SS                  |
|        +-| |---[00050  TON  T104 ]+
|         X00F
|
|-[R504]+-| |---------------------+---------------------[R505]
|        |R001                    |
|        |                        |
|        |BYP.SS                  |
|        +-| |---[00050  TON  T105 ]+
|         X00F
|
|-[R505]+-| |---------------------+---------------------[R506]
|        |X006                    |
|        |                        |
|        |BYP.SS                  |
|        +-| |---[00050  TON  T106 ]+
|         X00F
```

nection. The other four leads are always labeled 1, 2, 4, and 8. They connect to four separate terminals of a digital input module of the PLC.

Some BCD thumbwheel switches can be purchased with four or six digits. Others can be purchased in segments so that any number of these switches can be ganged together to form numbers with one, two, three, or more digits. In other words, if four sections of BCD thumbwheel switches are ganged together, the maximum number that can be loaded

Figure 6.11 Step Sequencer Circuit—Modified (Page 2)

```
|
|-[R506]+[00100   TON    T001 ]-------+------------------------( )-
|       |                             |                       R000
|       |                             |
|       |BYP.SS                       |
|       +-| |---[00050   TON   T107 ]+
|        X00F
|
|       +-----------+
|-| |---C   CNT     Q------------------------------------------( )-
|X005   |           |                                         R001
|       |           |
|       |           |
|-|/|---E 00005 C00 |
|R000   +-----------+
|
|-| |---------+------------------------------------------------( )-
|R501         |                                               Y020
|             |
|-| |---------+
|R503         |
|             |
|-| |---------+
|R504         |
|             |
|-| |----| |--+
|R506   R622
|
|-| |---------+------------------------------------------------( )-
|R502         |                                               Y021
|             |
|-| |---------+
|R505         |
|             |
|-| |----| |--+
|R506   R622
|
|-| |---------+------------------------------------------------( )-
|R503         |                                               Y022
|             |
|-| |---------+
|R505         |
|             |
|-| |----| |--+
|R506   R622
```

in is 9999. There are sixteen wires to connect to the input module of the PLC, along with the common connection.

Therefore, a four-section thumbwheel switch requires the complete use of a sixteen-point input module. Even if you plan to use only a single section of a BCD thumbwheel switch, which uses only four input terminals, you need to reserve one complete input module for the BCD thumbwheel switch. The reason is that the sixteen-point input modules

Figure 6.12 Step Sequencer Circuit—Modified (Page 3)

```
|
|-| |---------+---------------------------------------------------------( )-
|R502         |                                                        Y023
|             |
|             |
|-| |----| |--+
|R506    R622
|
|
|-| |---------+---------------------------------------------------------( )-
|R503         |                                                        Y024
|             |
|             |
|-| |---------+
|R505         |
|             |
|             |
|-| |----| |--+
|R506    R622
|
|
|-| |---------+---------------------------------------------------------( )-
|R502         |                                                        Y025
|             |
|             |
|-| |---------+
|R504         |
|             |
|             |
|-| |---------+
|R505         |
|             |
|             |
|-| |----| |--+
|R506    R622
|
| +----+
|-| END|
| +----+
```

are assigned a register of one word. Adding an additional push button or two to this particular input module is read by the register as part of the word. The register reads the sixteen-point input module as if an additional section of a thumbwheel switch had been added. The result is a considerably higher number than was set in the BCD thumbwheel switch.

The first circuit is called "Load Data from a Thumbwheel Switch to a Counter Circuit." It is designed to take a number set in a BCD thumbwheel switch, load it into an internal counter (C05), which operates an output coil (Y029), and load it into an external liquid crystal display (LCD) counter.

Inputs X000, X001, and X002 are connected to an input module in slot 1. The BCD thumbwheel switch is connected to the input module in slot 2 and therefore is assigned register XW01. The LCD counter is connected to the output module in slot 3 and therefore is assigned register YW02.

This circuit uses two function blocks to help attain the results. The first is FUN. NO. 50. It converts the number in the BCD thumbwheel switch (register XW01) into a binary number and stores it in a data register (in this example, D0100). The second function block is FUN. NO. 51. It converts the binary data in the internal counter C05 into a four-digit BCD code and loads it into register YW02 for the LCD counter.

When START input X000 is turned on, the number set on the thumbwheel switch is read into register XW01. This number is converted by FUN. NO. 50 from BCD to binary and is stored in data register D0100.

The number stored in data register D0100 is used as the preset for counter C05. When the preset is reached by the repeated operation of COUNT input X001, counter C05 turns on DIVERT output Y029. The current count value (not the preset value) in counter C05 is also converted to BCD by FUN. NO. 51 and is stored in register YW02, which is connected to the LCD counter for a continuously updated readout of the current count.

Figure 6.13 shows the first BCD thumbwheel circuit.

The second circuit is called "Select Recipes from a Thumbwheel Switch Circuit." It is designed to take the number set in a BCD thumb-

Figure 6.13 Load Data from a Thumbwheel Switch to a Counter Circuit

```
|START          THBWHL
|-| |----|^|---[XW01    BIN  D0100]
|X000   R00A
|
|COUNT  +-----------+                                   DIVERT
|-| |---C  CNT      Q---------------------------------------( )-
|X001   |           |                                   Y029
|       |           |
|RESET  |           |
|-|/|---E D0100 C05 |
|X002   +-----------+
|
|COUNT                DISPLY
|-| |---[C05     BCD1  YW02 ]
|X001
|
| +----+
|-| END|
| +----+
```

wheel switch, compare that number with a series of preset process recipe numbers, start up the process for the particular recipe that corresponds to the number in the BCD thumbwheel switch, and turn on the corresponding process pilot light.

The BCD thumbwheel switch is connected to the input module in slot 1; therefore, it is assigned register XW00. Input X015 is the RUN SS selector switch. It is connected to the input module in slot 2. The five pilot lights are connected to the output module in slot 3.

This circuit uses two types of function blocks to help attain the results. The first is FUN. NO. 50. It converts the number in the BCD thumbwheel switch (register XW00) into a binary number and stores it in a data register (in this example, D0200). The second type is FUN. NO. 25, which is the Constant Comparison Equals function block. Five of these FUN. NO. 25 function blocks compare the binary data in data register D0200 with the constants one, two, three, four, and five. These constants correspond to the process recipes numbered from one to five, which have been assigned to the corresponding numbers on the BCD thumbwheel switch.

When the RUN SS input X015 is turned on, the number set on the thumbwheel switch is read into register XW00. This number is converted by FUN. NO. 50 from BCD to binary and is stored in data register D0200.

The number stored in data register D0200 is compared by each FUN. NO. 25 to determine which recipe has been selected. For example, if the number *2* were set in the BCD thumbwheel switch, the comparison allows the start-up of the second recipe along with the pilot light labeled SYS 2 whenever the RUN SS selector switch is turned on.

This circuit does not contain the complete process recipe but simply provides the means to start up the process through the use of internal relays R021 through R025.

The benefit of a circuit like this is that all of your recipes can be programmed into one circuit for a machine or system, and can be started easily by simply setting the recipe number on the BCD thumbwheel switch. In other words, you do not have to shut down your machine to unload the current program before loading in the next program whenever a new process recipe is required. The second BCD thumbwheel circuit is shown in Figure 6.14.

One final note applies to users of the M20 and M40 PLCs. If you are using an M20 PLC without any expansion units in your operation, the largest BCD thumbwheel switch that you can use is one with three digits. In that case, the BCD thumbwheel switch would have to be the only input device used in the circuit.

Figure 6.14 Select Recipes from a Thumbwheel Switch Circuit

```
|RUN SS THBWHL
|-| |---[XW00    BIN  D0200]
|X015
|
|RUN SS                                                  SYS 1
|-| |--+[D0200  = .  00001]------------------------+--------( )-
|X015  |                                           |     R021
|      |                                           |
|      |                                           |     SYS1PL
|      |                                           +--------( )-
|      |                                                 Y021
|      |                                                 SYS 2
|      +[D0200  = .  00002]------------------------+--------( )-
|      |                                           |     R022
|      |                                           |
|      |                                           |     SYS2PL
|      |                                           +--------( )-
|      |                                                 Y022
|      |                                                 SYS 3
|      +[D0200  = .  00003]------------------------+--------( )-
|      |                                           |     R023
|      |                                           |
|      |                                           |     SYS3PL
|      |                                           +--------( )-
|      |                                                 Y023
|      |                                                 SYS 4
|      +[D0200  = .  00004]------------------------+--------( )-
|      |                                           |     R024
|      |                                           |
|      |                                           |     SYS4PL
|      |                                           +--------( )-
|      |                                                 Y024
|      |                                                 SYS 5
|      +[D0200  = .  00005]------------------------+--------( )-
|                                                  |     R025
|                                                  |
|                                                  |     SYS5PL
|                                                  +--------( )-
|                                                        Y025
|
| +----+
|-| END|
| +----+
```

If you are using an M40 PLC without any expansion units in your operation, you can use a two-digit, a three-digit, or a four-digit BCD thumbwheel switch in a circuit. A two-digit BCD thumbwheel switch can be connected to terminals X10–X17, which leaves the sixteen X0–XF terminals available for other input devices. Or a three-digit or a four-digit BCD thumbwheel switch can be connected to terminals X0–XF, which leaves the eight X10–X17 terminals available for other input devices.

SET POINT PROGRAM

This section presents a circuit that uses six different function blocks to create a Set Point Program. The purpose of this circuit is to compare a flow meter analog input signal with a preselected set-point value, and to control a variable frequency drive (VFD) pump through an analog output signal. This analog output signal increases or decreases the speed of the pump that controls the flow rate measured by the flow meter.

The function blocks used in this circuit to achieve these results are as follows:

FUN. NO. 14	Register Comparison (Greater than)
FUN. NO. 15	Register Comparison (Equal to)
FUN. NO. 16	Register Comparison (Less than)
FUN. NO. 20	Constant Addition
FUN. NO. 21	Constant Subtraction
FUN. NO. 50	Binary Conversion (BCD to Binary)

The PLC used in this circuit is an EX500. The modules are installed according to the I/O Allocation chart in Table 6.12.

In this circuit, the BCD thumbwheel switch is connected to the input module in slot 1. For more information on BCD thumbwheel switches, refer to the previous section.

The BCD thumbwheel switch and its register XW01 are used in conjunction with the Binary Conversion (BCD to Binary) function block

Table 6.12 EX500 I/O Allocation Chart

Slot No.	*0*	*1*	*2*	*3*	*4*
Module	Input (16 pt)	Input (16 pt)	Output (32 pt)	Input (Analog)	Output (Analog)
Register No.	XW00	XW01	YW02 YW03	XW04 XW05 XW06 XW07	YW08 YW09
Address No.	X000 thru X00F	XW01	Y020 thru Y02F and Y030 thru Y03F	XW04 XW05 XW06 XW07	YW08 YW09

(FUN. NO. 50) and the on-delay timer T120 and the off-delay timer T121 to set a clock rate. Notice that the two timers measure time in hundredths of a second, and the time value in both of the timers is taken from data register D0020. This is the register that stores the data from the BCD thumbwheel switch as a result of the operation of the Binary Conversion (BCD to Binary) function block (FUN. NO. 50). This means that the on and the off times are equal to each other and can be set by the operator.

This setting controls the CLOCK relay (R000), which controls the rate at which the comparison between the flow meter analog input signal and the selected set-point value is made, which is also when adjustments are made to the VFD analog output.

The main part of this circuit consists of a series of three comparisons between the FLOW analog input signal from the flow meter in register XW04 and the SET PT value stored in register XW05. The first comparison is on rung three and is made by the Register Comparison (Equal To) function block (FUN. NO. 15). If the digital value of the analog input signal in flow meter register XW04 and the digital value in set-point register XW05 are equal, CYCLE relay R001 is energized, which opens the normally closed (NC) R001 contact in the fourth rung so that the other two comparisons cannot be made. When the digital values in XW04 and XW05 are no longer equal, the CYCLE relay R001 is de-energized, and the other two comparisons can be executed.

The two function block comparisons on the fourth and fifth rungs are controlled by a NO contact of the CLOCK relay R000, an NC contact of the CYCLE relay R001, and the rising edge transitional contact R601. In other words, the two comparisons are made at the timed interval set in the BCD thumbwheel switch, but only if the Register Comparison (Equal to) function block (FUN. NO. 15) says that the two values are different. The transitional contact R601 further limits the comparison so that the comparison is made only at the first scan after the NO contact of the clock relay closes. This ensures that the comparisons are made with only one set of data and not with a series of numbers that change as the analog input changes with changing pump speeds and flow rates during the time frame of the cycle.

For a more complete discussion on the digital values for analog input modules, refer to the section entitled "Analog I/O."

The two other comparison function blocks are analyzed, but only one of them is actually energized. If it happens that the data in flow meter register XW04 is greater than the data in set-point register XW05, the Register Comparison (Greater than) function block (FUN. NO. 14) is energized.

In the fourth rung, the Constant Subtraction function block (FUN. NO. 21) has been connected directly to the Register Comparison (Greater than) function block (FUN. NO. 14). This means that FUN. NO. 14 begins to operate whenever FUN. NO. 21 is energized. The Constant Subtraction function block (FUN. NO. 21) subtracts a value of one (1) from the current digital value in the VFD analog output register YW08. This causes the pump to decrease in speed by the equivalent of one digital value, which reduces the flow rate and thus reduces the value sensed by the flow meter. This slight reduction in speed is held until the next scan, when the comparison process is repeated.

If it happens that the data in flow meter register XW04 is less than the data in set-point register XW05, the Register Comparison (Less than) function block (FUN. NO. 16) in rung five is energized.

In the fifth rung, the Constant Addition function block (FUN. NO. 20) is connected directly to the Register Comparison (Less than) function block (FUN. NO. 16). This means FUN. NO. 16 begins to operate whenever FUN. NO. 20 is energized. The Constant Addition function block (FUN. NO. 20) adds a value of one (1) to the current digital value in the VFD analog output register YW08. This causes the pump to increase in speed by the equivalent of one digital value, which increases the flow rate and thus increases the value sensed by the flow meter. This slight increase in speed is held until the next scan, when the comparison process is repeated. The circuit is shown in Figure 6.15.

Figure 6.15 Set Point Program

```
|
|START            THMWHL          TCYCLE
|-| |----|^|---[XW01     BIN  D0020]
|XOOF    R600
|
|CLOCK  TCYCLE          T-ON    TCYCLE          T-OFF              CLOCK
|-|/|---[D0020   TON    T120 ]-[D0020   TOF    T121 ]------------( )-
|R000                                                              R000
|
|FLOW          SET PT                                              CYCLE
|-[XW04     =    XW05 ]------------------------------------------( )-
|                                                                  R001
|
|CLOCK  CYCLE               FLOW           SET PT  VFD              VFD
|-| |----|/|----|^|--+[XW04       >     XW05 ]-[YW08   -. 1  >  YW08 ]
|R000    R001     R601 |
|                      |
|                      |FLOW           SET PT  VFD              VFD
|                      +[XW04       <     XW05 ]-[YW08   +. 1  >  YW08 ]
|
|
| +----+
|-| END|
| +----+
```

AVERAGING CIRCUIT

This section presents a circuit that uses three function blocks to arrive at a number that is the mathematical average of several numbers. The purpose of this circuit is to reduce the erratic action of an analog output caused by the fluttering of the analog input signal. Although no actual output is shown in this circuit, the output would use the data stored in data register D0170. The data in register D0170 is actually a single value that is the calculated average of the most recent fifty program scans of the analog input signal fed to register XW00.

The function blocks used are as follows:

FUN. NO. 0	Register Transfer (Word to Word)
FUN. NO. 5	Table Block Transfer (Table to Table)
FUN. NO. 64	Average Value

The circuit is shown in Figure 6.16.

At every scan of the program, the analog input signal is fed to register XW00. The Register Transfer function block (FUN. NO. 0) transfers the digital value of that data to register D0100.

The Table Block Transfer function block (FUN. NO. 5) creates two tables, each with fifty data registers. The first table of fifty registers starts with register D0100 and ends with register D0149. The second table of fifty registers starts with register D0101 and ends with register D0150. Note that the registers in the two tables overlap.

The data in that first table of fifty registers is transferred to the second table of fifty registers. The result of that transfer action is to move each piece of data down one register. In other words, the data from register D0100 in the first table goes to register D0101 in the second table; the

Figure 6.16 Averaging Circuit

```
|RUN
|-| |--+[XW00    W->W    D0100]
|R00A  |
|      |
|      |
|      +[D0100   T->T    [50]    D0101]
|      |
|      |
|      |
|      +[D0100    AVE    [50]    D0170]
|
|
| +----+
|-| END|
| +----+
```

data from register D0101 in the first table goes to register D0102 in the second table; the data from register D0102 in the first table goes to register D0103 in the second table; the data from register D0103 in the first table goes to register D0104 in the second table; and so on. The transfer leaves register D0100 open to receive the next scan of data.

The Average Value function block (FUN. NO. 64) calculates the average of the fifty numbers and transfers that average number to data register D0170. The number can be used elsewhere in the circuit just as the original XW00 register would have been used if the analog input signal had not been fluttering.

ANALOG I/O

The task of including analog inputs and analog outputs in your program is much easier than you are probably expecting it to be. This section describes analog inputs. Analog outputs are described in the next section, entitled "Function Generator Circuit."

Let's start with a basic example of a wet well that has a pressure sensor installed at the bottom for measuring the level of the water in the well. This sensor is connected to an analog input module in the PLC by means of a 1–5 VDC (or 4–20 ma or other low VDC) control voltage signal. This analog input signal can be used to control both analog and digital outputs. These input and output signals can also be used in conjunction with the other function blocks that are available to perform a variety of functions. For more information on the use of function blocks, refer to the earlier section, entitled "Function Block Programming."

This example uses the analog input signal from the pressure sensor to operate four pilot lights. The names, synonyms, and wet well levels chosen for activation of these pilot lights are as follows:

Name	*Synonym*	*Well Level*
Low Level shutdown	LLSD	1.0 ft
Low Level alarm	LLA	1.5 ft
High Level alarm	HLA	10.0 ft
High Level shutdown	HLSD	10.5 ft

To set up this circuit, the first task is to determine the voltage output of the pressure sensor relative to the depth of the water in the wet well. This can sometimes be accomplished with the data supplied by the sensor manufacturer, or it may have to be done by experimentation. For this particular example, the output of the sensor was 2.5 VDC when the depth of the water was 10.5 ft.

Toshiba has four analog input modules for the EX100 PLC and five analog input modules for the EX250 and the EX500 PLCs. These nine modules are assigned one of three different digital ranges (Table 6.13).

Note: The four EX100 analog input modules are the type that can be installed in an expansion rack for use with the M20 and the M40 PLCs.

This example illustrates Model No. AI21, which is an EX100 eight-bit module. This means we will be working with the digital range of 250.

The analog-digital (A/D) conversion chart in your PLC manual shows the correlation between the analog input value and the actual digital value to which the Toshiba module converts that analog value. For the A/D conversion chart for the M20 and the M40 PLCs, refer to page 45 of that PLC manual. For the A/D conversion chart for the EX100 PLC, refer to page 34 of that PLC manual. For the A/D conversion chart for the EX250 and the EX500 PLCs, refer to page 15 of the *EX250/500 Analog Input User's Manual.*

Working with the A/D conversion chart for the 1–5 VDC range for the AI21 analog input module and with the knowledge that the pressure sensor output is 2.5 VDC at the water level depth of 10.5 feet, you can construct an A/D conversion table for the example (Table 6.14).

For your projects, you can obtain the values needed to construct the A/D conversion table by one of two methods. The first method is to calculate the values. The second method is for those of you who can't stand the thought of making calculations. It requires that you make your own

Table 6.13 Analog Input Modules

PLC Type	*Model No.*	*No. of Channels*	*No. of Bits*	*Digital Range*
EX100	AI21	4	8	250
EX100	AI31	4	8	250
EX100	AI22	4	12	4000
EX100	AI32	4	12	4000
EX250	6290B20	2	11	±2000
EX250	6290B10	2	11	±2000
EX250	6290B5	2	11	±2000
EX250	6292V	8	11	±2000
EX250	6292C	8	12	±2000

Table 6.14 A/D Conversion Table

Synonym	*Well Level (ft)*	*VDC*	*Digital Value*
—	0.0	1.00	000
LLSD	1.0	1.14*	009*
LLA	1.5	1.21*	013*
HLA	10.0	2.43*	089*
HLSD	10.5	2.50†	094*
—	—	5.00	250

*Calculated value.
†Measured value.

chart and table. Your chart should be like the A/D conversion chart in the PLC manual, but it should be modified for your own use. Your chart should have the following five features:

1. It should be accurately drawn to scale.
2. It should be large enough so that you can accurately plot points on it.
3. It should include only the control voltage or amperage along the bottom line that you will be using in your operation.
4. It should include the digital values along the left-hand side that correspond to the analog input module that you will be using in your operation.
5. It should include the angled line for your voltage or amperage range that extends up to the maximum value.

When you have established the voltage output of the pressure sensor relative to the depth of the water in the wet well (or whatever else you are measuring), you can add another scale along the bottom line that represents the depth of the water level. In the example, the pressure sensor puts out a signal of 2.5 VDC when the depth of the water level is 10.5 feet.

Then select the depths that you want to establish as control points and lay them off to scale along that bottom line. In the example, the values are 1.0, 1.5, 10.0, and 10.5 feet of well level depth.

Now, extend the control point marks straight up until they intersect the angled line. Then draw horizontal lines from these intersection points over to the left-hand side of the chart. Then read the digital value for each of the control points where the horizontal lines intersect the vertical digital value line. These digital values can then be put into your

A/D conversion table and put into the function blocks that will be used in the circuit.

The circuit itself uses two different types of function blocks and four digital outputs (Figure 6.17).

The first two function blocks used are FUN NO. 26. They are the constant comparison in which the input data (XW00) is less than a constant.

The second two function blocks used are FUN NO. 24. They are the constant comparison in which the input data (XW00) is greater than a constant.

The value for the constant used for all of the function blocks is the calculated digital value for the water well depths as shown in the A/D Conversion Table that was developed earlier in this section. In both cases, the pilot light output is turned on whenever the result of the function block comparison is true.

These four alarm lights operate in the following way:

1. Whenever the water level is below 1.0 foot (the 00009 digital value), both the LLSD and the LLA pilot lights are on.
2. Whenever the water level is above 1.0 foot (the 00009 digital value) but below 1.5 feet (the 00013 digital value), only the LLA pilot light is on.
3. Whenever the water level is above 1.5 feet (the 00013 digital value) but below 10.0 feet (the 00089 digital value), no pilot lights are on.

Figure 6.17 Analog Alarm Circuit

```
|                                                         LLSD
|-[XW00    < . 00009]------------------------------------( )-
|                                                         Y05A
|
|                                                         LLA
|-[XW00    < . 00013]------------------------------------( )-
|                                                         Y05B
|
|                                                         HLA
|-[XW00    > . 00089]------------------------------------( )-
|                                                         Y05C
|
|                                                         HLSD
|-[XW00    > . 00094]------------------------------------( )-
|                                                         Y05D
|
| +-----+
|-| END |
| +-----+
```

4. Whenever the water level is above 10.0 feet (the 00089 digital value) but below 10.5 feet (the 00094 digital value), the HLA pilot light is on.
5. Whenever the water level is above the 10.5 feet (the 00094 digital value), both the HLA and the HLSD pilot lights are on.

As you have seen, the use of the analog features can be very easy. The most difficult task is determining the digital equivalent value for the sensor output voltage.

The next section presents a circuit that operates with both analog inputs and analog outputs in a unique manner.

FUNCTION GENERATOR CIRCUIT

The Function Generator function block (FUN. NO. 65) allows you to design a circuit with a special relationship between the analog input and the analog output. In other words, you can design a circuit that has a special or an unusual path for the analog output to follow relative to the analog input data. Let me briefly mention two uses of this feature. First, you can design a circuit in which the analog output decreases in a straight line manner as the analog input increases in a straight line manner. Second, you can design a circuit in which the analog output reacts in anything but a straight line manner as the analog input increases in a straight line manner.

This second use of this feature is particularly helpful when you have two or three pumps that are trying to keep a wet well from overflowing in a lead-lag or a lead-lag-last mode of operation. In the lead-lag operation, as the analog input sensor indicates that the level is rising, the pumps turn on in a sequential manner. However, with the use of the function generator, you can reduce the speed of the lead pump when you bring the lag pump on-line, so that the two pumps can operate at the same speed and are, therefore, sharing the load. You can do the same thing in a lead-lag-last operation. In this case, you can reduce the speed of both the lead pump and the lag pump when you bring the last pump on-line, so that all three of the pumps can operate at the same speed and are equally sharing the load.

The example shown in this chapter is for a circuit with a single analog input from a wet well level sensor and a single analog output, which represents the first of three VFD pumps.

The first step is to define the relationship between the analog input and the analog output. The best way to do this is by plotting a Function

Generator Graph. That leads to a Function Generator Data Table. The plot of the graph is shown in Figure 6.18.

To show the diversity of the function generator, we have designed the analog output to vary all over the map as the analog input signal continues to rise.

The numbers along the bottom of the graph (X) represent the digital values for the analog input signals. The digital values of 0–250 represent the fact that this circuit is being written around an EX100 PLC with a Model No. AI21 analog input module, which has been installed in slot 1. For a complete description about how to work with the digital values for analog inputs, refer to the previous section entitled "Analog I/O."

The numbers up the side of the graph (Y) represent the digital values for the analog output signals. The digital values of 0–250 represent the fact that this circuit is being written for the EX100 PLC with a Model No. AO31 analog output module, which has been installed in slot 2. The range of 0–250 also represents the full range of the pump speed.

Toshiba produces three analog output modules for the EX100 PLC

Figure 6.18 Function Generator Graph

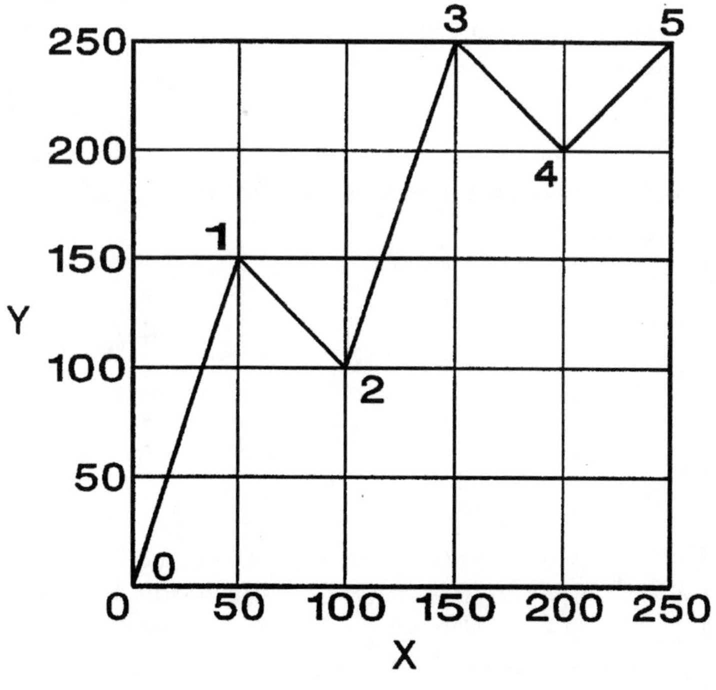

Table 6.15 Analog Output Modules

PLC Type	*Model No.*	*No. of Channels*	*No. of Bits*	*Digital Range*
EX100	AO31	2	8	250
EX100	AO22	2	12	4000
EX100	AO32	2	12	4000
EX250	6295B20	2	11	±2000
EX250	6295B10	2	11	±2000
EX250	6295B5	2	11	±2000
EX250	6295U20	2	12	±2000
EX250	6295U5	2	12	±2000

and five analog output modules for the EX250 and the EX500 PLCs. These eight modules are assigned one of three different digital ranges (Table 6.15).

Note: The three EX100 analog output modules are the type that can be installed in an expansion rack for use with the M20 and M40 PLCs.

All of the points on the Function Generator Graph at which there is a change of direction in the speed of the output must be accounted for, so the six points have been labeled as 0, 1, 2, 3, 4, and 5. From these six points, we have constructed Table 6.16.

Table 6.16 Function Generator Data Table

Component of Each Data Point	*Digital Value*	*Data Register No.*
X of 0	000	D0600
X of 1	050	D0601
X of 2	100	D0602
X of 3	150	D0603
X of 4	200	D0604
X of 5	250	D0605
Y of 0	000	D0606
Y of 1	150	D0607
Y of 2	100	D0608
Y of 3	250	D0609
Y of 4	200	D0610
Y of 5	250	D0611

Notice that each point has been broken down into its X (input) and Y (output) components. Therefore, the number of data registers actually required to perform this operation is double the number of points plotted on the graph. Note that the first half of the data registers contain the X components and the last half of the data registers contain the Y components. *This configuration is mandatory.*

Now these digital values for X and Y can be loaded into the data registers, according to the following steps:

1. Start at the TOP MENU. Press F7–Enter/Change program name. Overwrite the word *default* with a new program name, and press F10 to save that name and return to the TOP MENU.
2. Press F3–Block Monitor, which takes you to a block monitor screen that looks like Table 6.17.
3. Press F1–Start Number, and a message appears that reads as follows: **`Enter the new starting register number: YW00.`**
4. Overwrite the YW00 with **D0600**, and press ENTER.

Table 6.17 Initial Block Monitor for Design of Function Generator Circuit

Reg.	Value	FEDC	BA98	7654	3210	Reg.	Value	FEDC	BA98	7654	3210
YW00	00000	0000	0000	0000	0000	YW16	00000	0000	0000	0000	0000
YW01	00000	0000	0000	0000	0000	YW17	00000	0000	0000	0000	0000
YW02	00000	0000	0000	0000	0000	YW18	00000	0000	0000	0000	0000
YW03	00000	0000	0000	0000	0000	YW19	00000	0000	0000	0000	0000
YW04	00000	0000	0000	0000	0000	YW20	00000	0000	0000	0000	0000
YW05	00000	0000	0000	0000	0000	YW21	00000	0000	0000	0000	0000
YW06	00000	0000	0000	0000	0000	YW22	00000	0000	0000	0000	0000
YW07	00000	0000	0000	0000	0000	YW23	00000	0000	0000	0000	0000
YW08	00000	0000	0000	0000	0000	YW24	00000	0000	0000	0000	0000
YW09	00000	0000	0000	0000	0000	YW25	00000	0000	0000	0000	0000
YW10	00000	0000	0000	0000	0000	YW26	00000	0000	0000	0000	0000
YW11	00000	0000	0000	0000	0000	YW27	00000	0000	0000	0000	0000
YW12	00000	0000	0000	0000	0000	YW28	00000	0000	0000	0000	0000
YW13	00000	0000	0000	0000	0000	YW29	00000	0000	0000	0000	0000
YW14	00000	0000	0000	0000	0000	YW30	00000	0000	0000	0000	0000
YW15	00000	0000	0000	0000	0000	YW31	00000	0000	0000	0000	0000

5. The screen changes, and register D0600 is at the head of the column. The screen looks like Table 6.18.
6. Press F5–Set Value, and a message appears that reads as follows:

   `Register D0600 ... Enter New Value: 00000.`

7. Press ENTER, because the value that needs to be in that register is 00000.
8. Press F4–Inc & Set, and a message appears that reads as follows:

   `Register D0601 ... Enter New Value: 00000.`

9. Overwrite this value with **00050**, which is the value for D0601 from the Function Generator Data Table, and press ENTER.
10. A confirmation message appears near the bottom of the screen, and the number 00050 appears in the Value column. The value for 50 also appears in the appropriate register bit columns of F through 0.
11. Repeat steps 8 and 9 until all twelve data points have been loaded into the block monitor.
12. When all of the data has been loaded, the block monitor screen will look like Table 6.19.

Table 6.18 Block Monitor After Starting Register Number Is Entered

Reg.	Value	FEDC	BA98	7654	3210	Reg.	Value	FEDC	BA98	7654	3210
D0600	00000	0000	0000	0000	0000	D0616	00000	0000	0000	0000	0000
D0601	00000	0000	0000	0000	0000	D0617	00000	0000	0000	0000	0000
D0602	00000	0000	0000	0000	0000	D0618	00000	0000	0000	0000	0000
D0603	00000	0000	0000	0000	0000	D0619	00000	0000	0000	0000	0000
D0604	00000	0000	0000	0000	0000	D0620	00000	0000	0000	0000	0000
D0605	00000	0000	0000	0000	0000	D0621	00000	0000	0000	0000	0000
D0606	00000	0000	0000	0000	0000	D0622	00000	0000	0000	0000	0000
D0607	00000	0000	0000	0000	0000	D0623	00000	0000	0000	0000	0000
D0608	00000	0000	0000	0000	0000	D0624	00000	0000	0000	0000	0000
D0609	00000	0000	0000	0000	0000	D0625	00000	0000	0000	0000	0000
D0610	00000	0000	0000	0000	0000	D0626	00000	0000	0000	0000	0000
D0611	00000	0000	0000	0000	0000	D0627	00000	0000	0000	0000	0000
D0612	00000	0000	0000	0000	0000	D0628	00000	0000	0000	0000	0000
D0613	00000	0000	0000	0000	0000	D0629	00000	0000	0000	0000	0000
D0614	00000	0000	0000	0000	0000	D0630	00000	0000	0000	0000	0000
D0615	00000	0000	0000	0000	0000	D0631	00000	0000	0000	0000	0000

Table 6.19 Block Monitor After Data Has Been Loaded

Reg.	Value	FEDC	BA98	7654	3210	Reg.	Value	FEDC	BA98	7654	3210
D0600	00000	0000	0000	0000	0000	D0616	00000	0000	0000	0000	0000
D0601	00050	0000	0000	0011	0010	D0617	00000	0000	0000	0000	0000
D0602	00100	0000	0000	0110	0100	D0618	00000	0000	0000	0000	0000
D0603	00150	0000	0000	1001	0110	D0619	00000	0000	0000	0000	0000
D0604	00200	0000	0000	1100	1000	D0620	00000	0000	0000	0000	0000
D0605	00250	0000	0000	1111	1010	D0621	00000	0000	0000	0000	0000
D0606	00000	0000	0000	0000	0000	D0622	00000	0000	0000	0000	0000
D0607	00150	0000	0000	1001	0110	D0623	00000	0000	0000	0000	0000
D0608	00100	0000	0000	0110	0100	D0624	00000	0000	0000	0000	0000
D0609	00250	0000	0000	1111	1010	D0625	00000	0000	0000	0000	0000
D0610	00200	0000	0000	1100	1000	D0626	00000	0000	0000	0000	0000
D0611	00250	0000	0000	1111	1010	D0627	00000	0000	0000	0000	0000
D0612	00000	0000	0000	0000	0000	D0628	00000	0000	0000	0000	0000
D0613	00000	0000	0000	0000	0000	D0629	00000	0000	0000	0000	0000
D0614	00000	0000	0000	0000	0000	D0630	00000	0000	0000	0000	0000
D0615	00000	0000	0000	0000	0000	D0631	00000	0000	0000	0000	0000

13. To save this data into the registers and to return to the TOP MENU, press F10.

Refer to the section entitled "Function Block Programming" for the instructions on how to locate and bring up FUN. NO. 65 while in the EX-PDD program. The circuit can be programmed as shown in Figure 6.19.

This circuit, as simple-looking as it is, is amazingly powerful.

When the RUN R022 contact (from another location in the circuit) is closed, the function generator goes into operation. As long as contact

Figure 6.19 Function Generator Circuit

```
|RUN
|-| |---[XW00    FG    [06]    D0600      >    YW05 ]
|R022
|
| +----+
|-| END|
| +----+
```

R022 is closed, the analog data from the wet well level sensor is sent to register XW00 and is converted into the equivalent digital value with each scan of the program.

That digital value is compared with the X component values of the six data points loaded into the twelve data registers starting at D0600. The function generator will then send the corresponding Y component to register YW05 for the analog output module, and the speed of the pump is set accordingly.

The Function Generator Graph (Figure 6.18) shows that the value of the X component along the bottom line increases as the level of the wet well increases. The speed of the pump increases along the line from point 0 to point 1. However, as the well level continues to increase, the pump speed decreases from point 1 to point 2. The continued rise of the well level drives the pump to its maximum speed at point 3. A continued rise in the well level drops the pump speed to 80 percent of its maximum speed (the 200 digital value divided by the 250 digital value) at point 4 before taking the pump back up to its maximum speed at point 5. As the well level decreases, the pump speed follows the same path from point 5 to point 0.

The six points on this function generator graph for this circuit represent a variation of a real situation. They have been modified to show the versatility of the function generator.

7 CHANGING PLCs

REPLACING A SMALL EX-STYLE PLC WITH AN M20 OR M40 PLC

You will often find it desirable or necessary to replace an existing Small Toshiba programmable logic controller (PLC) (an EX14B, EX28B, EX20, EX20+, EX40, EX40+, or EX40H) with an M20 or M40 PLC. In a large number of these cases, you will be able to determine that the relay ladder logic (RLL) in the current PLC is also the circuit that you want to use in the new PLC.

However, a problem immediately presents itself. The EX-PDD2040 software program used with the Small PLCs operates differently than the EX-PDD250 program used with the Medium PLCs. The EX-PDD250 treats some of the devices in a manner different from the EX-PDD2040. The Small PLCs also use octal addresses, whereas the Medium PLCs use hexadecimal addresses.

There are two different ways to resolve this problem. The first method can be used for short RLL circuits. It involves rewriting the complete circuit into the EX-PDD250 program using a print-out of the EX-PDD2040 circuit as a guide. Although this method can be quite useful for circuits with only a few pages, it becomes tedious for circuits of great length.

However, there is a second method for longer circuits, and it involves some copy work and some conversion work. The copy work can be performed as a DOS task with your personal computer (PC). The files from the circuit in the EX-PDD2040 program are copied (by means of DOS commands) to a new file (or subdirectory) in the EX-PDD250 program. The reason for copying the files is that some of the devices transfer directly between the two programs, and that saves a lot of work. Devices that do not transfer directly must be converted to the EX-PDD250 format. That conversion work is the subject of this chapter.

The best place to begin a conversion is with a review of the current RLL in the EX-PDD2040 program. The most important aspect of this review is to make sure that *all* input, output, and internal relay devices have been assigned synonyms. If not, now is the time to do it.

While you're at it, turn over a new leaf, make a New Year's resolution, and make the commitment to add synonyms to all of your future circuits as you work on them.

When all devices in the current program have been assigned synonyms, go to the F4–Documentation portion of the software and print a copy of the circuit. Be sure to include the Device Commentary Synonym and the Device Commentary (Synonym/Description) portions in the print-out. Better yet, make two copies. Keep one for your records and use the second one as your working copy.

COPYING

To copy this revised file over to the EX-PDD250 software, start by going to your EX-PDD250 directory and creating a subdirectory for the new operation. Let's say that the operation being modified is called the Line Tensioner. For the sake of brevity, and because of DOS limitations, we will call this subdirectory TENSION.

To create this subdirectory, go to the DOS prompt and change to the EX-PDD250 directory. Refer to the section entitled "EXPDD Software Quick Start" in chapter 2 for a short discussion of EX-PDD directory names. Your DOS prompt should read as follows:

```
C:\PDD-250>
```

To create the new subdirectory, type the following after the DOS prompt:

```
md tension
```

Press ENTER, and you now have a subdirectory in the EX-PDD250 directory named TENSION.

Copy all of the files from the Line Tensioner circuit in the EX-PDD2040 directory into the EX-PDD250\TENSION subdirectory by using DOS commands.

When you went to the F4-Documentation portion of the software and selected the Device Commentary (Synonym/Description) portion, the print-out included a sheet entitled "Device Number Comments."

Look at that Device Number Comments sheet and review all of the inputs that are listed. Because the Toshiba Small PLCs use the octal numbering system, and the Toshiba Medium PLCs use the hexadecimal

Table 7.1 Inputs for EX20 or EX20+ PLCs to M20 PLC—Basic Units

Original EX20/20+ Octal Address	*New M20 Hexadecimal Address*	*Comments*
X000	X000	OK as is
X001	X001	OK as is
X002	X002	OK as is
X003	X003	OK as is
X004	X004	OK as is
X005	X005	OK as is
X006	X006	OK as is
X007	X007	OK as is
X010	X008	Change
X011	X009	Change
X012	X00A	Change
X013	X00B	Change

numbering system, you will have to convert all of the input addresses from octal to hexadecimal.

Review the list of inputs in the Device column of your "Device Number Comments" sheet and cross out all of the original octal addresses. Pencil in the new hexadecimal addresses for each input according to Table 7.1, 7.2, 7.3, or 7.4.

Table 7.2 Inputs for EX20 or EX20+ PLCs to M20 PLC—Expansion Units

Original EX20/20+ Octal Address	*New M20 Hexadecimal Address*	*Comments*
X020	X030	Change
X021	X031	Change
X022	X032	Change
X023	X033	Change
X024	X034	Change
X025	X035	Change
X026	X036	Change
X027	X037	Change
X030	X038	Change
X031	X039	Change
X032	X03A	Change
X033	X03B	Change

Table 7.3 Inputs for EX40 or EX40+ PLCs to M40 PLC—Basic Units

Original EX40/40+ Octal Address	*New M40 Hexadecimal Address*	*Comments*
X000	X000	OK as is
X001	X001	OK as is
X002	X002	OK as is
X003	X003	OK as is
X004	X004	OK as is
X005	X005	OK as is
X006	X006	OK as is
X007	X007	OK as is
X010	X008	Change
X011	X009	Change
X012	X00A	Change
X013	X00B	Change
X014	X00C	Change
X015	X00D	Change
X016	X00E	Change
X017	X00F	Change
X020	X010	Change
X021	X011	Change
X022	X012	Change
X023	X013	Change
X024	X014	Change
X025	X015	Change
X026	X016	Change
X027	X017	Change

Now it's time to review that "Device Number Comment" sheet again. Review the list of outputs in the *Device* column of the Device Number Comments sheet and cross out all of the original octal addresses. Pencil in the new hexadecimal addresses for each output according to Table 7.5, 7.6, 7.7, or 7.8.

As you went through these address conversion procedures, it should have quickly become obvious that the only thing that is tying the new hexadecimal address back to the original octal address is the synonym. Therefore, the importance of assigning synonyms to every input and output should now be apparent.

Assigning synonyms to the internal relays (R000, etc.) also is helpful, but it is not nearly as important as it is for the input and output devices. Leave the addresses of the internal relays as they are. They are in the oc-

Table 7.4 Inputs for EX40 or EX40+ PLCs to M40 PLC—Expansion Units

Original EX40/40+ Octal Address	*New M40 Hexadecimal Address*	*Comments*
X030	X030	OK as is
X031	X031	OK as is
X032	X032	OK as is
X033	X033	OK as is
X034	X034	OK as is
X035	X035	OK as is
X036	X036	OK as is
X037	X037	OK as is
X040	X038	Change
X041	X039	Change
X042	X03A	Change
X043	X03B	Change
X044	X03C	Change
X045	X03D	Change
X046	X03E	Change
X047	X03F	Change
X050	X040	Change
X051	X041	Change
X052	X042	Change
X053	X043	Change
X054	X044	Change
X055	X045	Change
X056	X046	Change
X057	X047	Change

tal system, but changing them to hexadecimal creates more work and does not provide any real benefits.

CONVERTING

Now it's time to see what your copied circuit looks like on the screen and to start the work of converting the circuit. Start the EX-PDD250 program, and proceed as follows:

1. From the TOP MENU, select F7–Enter/Change program name.
2. While at the Enter/Change program name screen, type over the word *default* with the name of your program.

Table 7.5 Outputs for EX20 or EX20+ PLCs to M20 PLC—Basic Units

Original EX20/20+ Octal Address	*New M20 Hexadecimal Address*	*Comments*
Y000	Y020	Change
Y001	Y021	Change
Y002	Y022	Change
Y003	Y023	Change
Y004	Y024	Change
Y005	Y025	Change
Y006	Y026	Change
Y007	Y027	Change

3. Press F10. This saves the name and brings you back to the TOP MENU.
4. Press F2–Ladder Logic.

It won't take long to recognize that this circuit is in trouble. Many of the devices are incomplete, and that's what this conversion process is all about.

In the process of converting a circuit, each device must be reviewed separately to see what, if anything, must be done to properly complete the conversion.

Table 7.6 Outputs for EX20 or EX20+ PLCs to M20 PLC—Expansion Units

Original EX20/20+ Octal Address	*New M20 Hexadecimal Address*	*Comments*
Y010	Y050	Change
Y011	Y051	Change
Y012	Y052	Change
Y013	Y053	Change
Y014	Y054	Change
Y015	Y055	Change
Y016	Y056	Change
Y017	Y057	Change

Table 7.7 Outputs for EX40 or EX40+ PLCs to M40 PLC—Basic Units

Original EX40/40+ Octal Address	*New M40 Hexadecimal Address*	*Comments*
Y000	Y020	Change
Y001	Y021	Change
Y002	Y022	Change
Y003	Y023	Change
Y004	Y024	Change
Y005	Y025	Change
Y006	Y026	Change
Y007	Y027	Change
Y010	Y028	Change
Y011	Y029	Change
Y012	Y02A	Change
Y013	Y02B	Change
Y014	Y02C	Change
Y015	Y02D	Change
Y016	Y02E	Change
Y017	Y02F	Change

Table 7.8 Outputs for EX40 or EX40+ PLCs to M40 PLC—Expansion Units

Original EX40/40+ Octal Address	*New M40 Hexadecimal Address*	*Comments*
Y020	Y050	Change
Y021	Y051	Change
Y022	Y052	Change
Y023	Y053	Change
Y024	Y054	Change
Y025	Y055	Change
Y026	Y056	Change
Y027	Y057	Change
Y030	Y058	Change
Y031	Y059	Change
Y032	Y05A	Change
Y033	Y05B	Change
Y034	Y05C	Change
Y035	Y05D	Change
Y036	Y05E	Change
Y037	Y05F	Change

Table 7.9 summarizes each of the devices and the changes that are necessary to convert them from the EX-PDD2040 format to the EX-PDD250 format. A complete description of the conversion of each of these devices follows.

COILS

The coils for internal relays and outputs both transfer fairly well. There are only two minor changes that need to be made.

The result of the conversion places the coils one column to the left of where they need to be. Work on one coil at a time, and insert a new coil in the column to the right of the original coil. Refer to your revised "Device Number Comments" sheet, and assign the new address to this new coil.

The synonym is inserted along with the new address if the new coil is an internal relay that has kept its original address. However, if the address has changed, no synonym will be inserted, because the synonym is still associated with the original address. So the address assigned to the synonym must be changed. To do this, press Shift-F9–Assign Device to bring the three-line REF/SYN chart to the bottom of the screen. Then

Table 7.9 Changes Necessary to Convert from EX-PDD2040 Format to EX-PDD250 Format

Device	*Conversion Instruction Summary*
ADDRESSES	All input and output addresses must be converted.
COILS	Must be changed slightly.
CONTACTS	OK as converted, except for timers and counters.
COUNTERS	Must be changed completely.
END STATEMENT	Must be replaced.
FLIP/FLOPS	Must be changed completely.
JCS & JCR	Recommend that they be replaced.
MCS & MCR	Recommend that they be replaced.
MISCELLANEOUS	More rungs can be added to each page.
SHIFT REGISTERS	Must be changed completely.
STEP SEQUENCERS	Must be changed completely.
SYNONYMS	OK as converted; all I/O *must* be assigned synonyms.
SYSTEM and CONTROL INFORMATION	The PLC type must be selected and the I/O Allocation must be performed.
TIMERS	Must be changed completely.
TRANSITIONAL CONTACTS	Recommend that they be replaced.
VALIDATE	Must be done.

press F9–Toggle Ref/syn. The SYN column becomes the controlling column. The word *SYN* at the top of the column is highlighted, and the cursor is moved to that column. Move the cursor down to your coil synonym and then move the cursor back to the REF column. Type over the original octal address with the new hexadecimal address, and press ENTER. Press F10 to return to the circuit. Your new coil will now have the synonym, and the original coil will not.

Delete the original coil by pressing F4, which replaces it with a solid line.

CONTACTS

The normally open (NO) and normally closed (NC) contacts of inputs, outputs, and internal relays all transfer over to the new program without any problems. All you have to do is assign the input and output contacts their new addresses and synonyms. To do that, refer to your revised "Device Number Comments" sheet, and assign the new addresses to these contacts.

The synonym is inserted with the new address if the contact is from an input or an internal relay that has kept its original address.

However, if the address has changed, no synonym will be inserted, because the synonym is still associated with the original address. The address assigned to the synonym must be changed.

To change the synonym, press Shift-F9–Assign Device to bring the three-line REF/SYN chart to the bottom of the screen. Then press F9–Toggle Ref/syn. The SYN column becomes the controlling column. The word *SYN* at the top of the column is highlighted, and the cursor is moved to that column. Move the cursor down to your contact synonym and then move the cursor back to the REF column. Type over the original octal address with the new hexadecimal address, and press ENTER. Press F10 to return to the circuit. Your contact will now have the synonym.

However, there are two exceptions to the statement about contacts in the first paragraph. The EX-PDD250 software does not recognize the NO or the NC contacts that are associated with timers and counters. For more information on the conversion of counters and timers, refer to the following sections on counters and timers.

COUNTERS

In the EX-PDD2040 software, counters are designed to be configured like counting relays. Therefore, they are composed of coils and contacts. The coil portion is automatically inserted at the right-hand end of the rung. The contacts are then inserted into the circuit wherever they are needed.

However, in the EX-PDD250 software, counters are configured as complete devices and can be installed anywhere in the rung, except at the right-hand end. Therefore, the output from a counter now leads directly to an output coil. If the output from that counter needs to be used in several rungs in the circuit, the output should be set up as the coil of an internal relay, with contacts from that relay being used in the other rungs. In other words, the original counter coil positions must be deleted from the right-hand side of the rung, and the new counter device must be inserted in front of the coil or the other device that the counter was controlling in the original circuit.

If the original counter used only one NO contact in the circuit, replace that NO contact directly with the new counter device.

If the original counter used more than one NO contact, or if there were any NC contacts in the circuit, replace all them with NO and NC contacts from the newly created internal relay that is now controlled by the output of the counter. Assign a synonym to this new internal relay that has a direct relationship to the counter. For example, if the counter address is C05, assign a synonym such as CNT05 to the new internal relay for ease of recognition during circuit troubleshooting.

For a more complete discussion about counters, refer to the section entitled "Counters" in chapter 3.

END STATEMENT

The END statement must be replaced. Simply delete the original END statement, and insert a new END statement.

FLIP/FLOPS

The commands for the flip/flops in the EX-PDD2040 software are handled completely differently from those in the EX-PDD250 software, and they do not transfer between the two programs. As a result, during the conversion operation, all of the original flip/flops must be deleted, and new ones must be inserted. For details on these devices, refer to the section entitled "Flip/flops" in chapter 3.

JCS AND JCR

It is recommended that the Jump Control Set (JCS) and the Jump Control Reset (JCR) commands be replaced with the commands from the EX-PDD250 software.

MCS AND MCR

It is recommended that the Master Control Set (MCS) and the Master Control Reset (MCR) commands be replaced with the commands from the EX-PDD250 software.

MISCELLANEOUS

The number of pages of a converted circuit can be reduced by nearly one-half. The EX-PDD2040 software allows a circuit to have eight rungs per page, whereas the EX-PDD250 software allows fourteen rungs. Rungs from the top of one page can be transferred to the bottom of the preceding page in a systematic method to reduce the page count. This is accomplished with the Move/Copy command, which can be accessed through either the F3–Global Page or the F4–Edit Page commands. Refer to the *EX-PDD Software User's Manual* for more information on this subject.

SHIFT REGISTERS

The commands for the shift registers in the EX-PDD2040 software are handled completely differently from those in the EX-PDD250 software, and they do not transfer between the two programs. As a result, all of the original shift registers must be deleted, and new ones must be inserted. For details on these devices, refer to the section entitled "Shift Registers" in chapter 3.

STEP SEQUENCERS

The STIN and STOT commands stand for the step input and the step output commands for a step sequencer in the EX-PDD2040 software. The step sequencer in the EX-PDD250 software contains three commands that are handled completely differently from the STIN and STOT commands. As a result, all of the STIN and STOT commands must be deleted, and the new commands from the EX-PDD250 software must be inserted. For details on these three devices, refer to the section entitled "Step Sequencer" in chapter 6.

SYNONYMS

The synonyms transfer between the two EX-PDD formats without any problems. Remember that the synonyms complete the bridge between the two software formats whenever there is an address change. Therefore, all devices that have had addresses changed must also have the synonyms changed to the new addresses. Refer to the sections in this chapter entitled "coils and "contacts" for more information on this subject.

SYSTEM and CONTROL INFORMATION

The type of PLC with which you are working must be assigned, and the Input-Output (I/O) Allocation operation must be performed. Go to the TOP MENU, and press F1–System and Control Information.

To assign the PLC type, press F1–System Information. The screen opens with the cursor at the Program ID line; this is where you type the name of your program. Move the cursor down to the PLC Type line. Choose the type of PLC to which you are converting the circuit by pressing the space bar at the bottomof the keyboard until your type of PLC is presented. The PLC choices are as follows: M20/M40, EX100, EX200B, EX250 with 2K steps, EX250 with 4K steps, and EX500. When you have completed this selection process, press F10 twice to return to the TOP MENU.

You must perform the I/O Allocation operation before you attempt to validate the converted circuit and before you attempt to load the program into the PLC. The subject of I/O Allocation is described in the section entitled "Automatic Identification of the I/O Allocation" in chapter 4, and in the section entitled "I/O Allocation" in chapter 6.

TIMERS

In the EX-PDD2040 software, timers are designed to be configured like timing relays. Therefore, they are composed of coils and contacts. The coil portion is automatically inserted at the right-hand end of the rung. The contacts are then inserted into the circuit wherever they are needed.

However, in the EX-PDD250 software, timers are configured as complete devices and can be installed anywhere in the rung, except at the right-hand end. Therefore, the output from a timer now leads directly to an output coil. If the output from that timer needs to be used in several rungs in the circuit, the output should be set up as the coil of an internal relay, with contacts from that relay being used in the other rungs.

In other words, the original timer coil portion must be deleted from the right-hand side of the rung, and the new timer device must be inserted in front of the coil or the other device that the timer was controlling in the original circuit.

If the original timer used only one NO contact in the circuit, replace that NO contact directly with the new timer device. If the original timer used more than one NO contact, or if there were any NC contacts in the circuit, replace all of them with NO and NC contacts from the newly created internal relay that is now controlled by the output of the timer. Assign a synonym to this new internal relay that has a direct relationship to the timer. For example, if the timer address is T030, assign a synonym such as TIM030 to the new internal relay for ease of recognition during circuit troubleshooting. For a more complete discussion of timers, refer to the section entitled "Timers" in chapter 3.

TRANSITIONAL CONTACTS

It is recommended that the original transitional contacts be replaced with new transitional contacts from the EX-PDD250 software.

VALIDATE

After you have completed the conversion process, the circuit must be validated. The VALIDATE routine checks the circuit for gross errors only. It checks for devices that do not have addresses, for openings in rungs, for short circuits, for vertical lines that are not connected, for pages that contain too many instructions, and for other errors that the software does not allow you to enter into the PLC.

VALIDATE does not tell you if your logic is correct. It does not tell you if your circuit will operate as you intended to program it. It does not tell you that the timer you wanted set for thirty seconds will time out at three seconds. Nor does it tell you that you have programmed your largest motor to be cycling on and off every two seconds.

To validate any single page, press F7–Valid Ladder at command line 1. If you press Shift-F7–Valid Ladder, the program checks each page sequentially, and if all is well, you will receive the following message:

```
Validating Ladder Logic Complete. XYZ Steps Used.
```

If you have made an error that VALIDATE does not accept, the validation process is stopped, and the program takes you to the page that contains the first problem. It puts the program into the EDIT mode and places the cursor at the point of the error. All you have to do is recognize the nature of the error and correct it. Make the correction, ACCEPT the page, and start the Shift-F7–Valid Ladder routine again. Repeat this process until the following message appears:

```
Validating Ladder Logic Complete. XYZ Steps Used.
```

You have completed the process of converting a circuit that was written in the EX-PDD2040 software into a new circuit in the EX-PDD250 software. The only thing left to do is to test the circuit.

8 A BAKER'S DOZEN PLC TIPS

CIRCUIT PROTECTION

1. When designing your control system, provide a fuse or circuit breaker for the 120 VAC line that feeds power to the PLC.
2. When designing your control system, also provide an extra fuse or circuit breaker that can handle *all* of the inputs.
3. When designing your control system, and when sizing your control panel, provide a fuse or circuit breaker for *each* of the outputs. Providing protection for each output at the start of the project can save many hours of troubleshooting in the future, especially when you try to find the source of a blown fuse if eight, ten, or even more outputs are on a single fuse. One excellent method of accomplishing this protection is to install terminal strips that have built-in fuse holders.

PROGRAMMING

4. Are you still using a Toshiba programmer for programming and troubleshooting your PLC circuits? Consider trying a personal computer (PC) or laptop computer with the EX-PDD software that has been designed for your PLC. Both programming and troubleshooting will become more enjoyable and more efficient.
5. Use transitional contacts very carefully and sparingly. They can create more problems than they solve, unless you are extremely proficient with their use and have determined that a transitional contact is absolutely necessary. I saw one system that was designed to use a transitional contact and a bar-code reader to count the shipping cases on a slow-moving packaging line. It was a bad scene—until the transitional contact was eliminated.

6. Hard-wire all inputs as normally open (NO) and close them with the software. In this way, you can establish a point of reference between the ladder logic programmer who designs the circuit and the electrician who has to hard-wire it.
7. PLC inputs can be used an unlimited number of times in your circuit. Instead of having to worry (as we did in the past) about the number of poles on a selector switch or a push button, only one contact on a switch needs to be wired to the PLC. The contact can be used in the circuit as a NO or normally closed (NC) input contact as many times as needed.
8. In contrast to the multiple use of inputs, as mentioned in Tip 7, outputs must be treated differently. A common cause of circuit problems is the attempted use of an output (for instance, Y020) in two or more lines in the circuit. Although this can be done under some very specific circumstances, it is best to limit your use of this output (Y020) to only one line in the entire circuit. If your system requires that output Y020 be used in several different sequences of the circuit, construct the inputs in a parallel configuration so that they all feed into the single Y020 output.
9. In the same way that an input can be used many times, so can the contacts of an internal PLC relay. Gone are the days when circuit designers had to count the number of times that they used the contacts of a hard-wired relay to make sure they had not exceeded the four Form C or the two NO and two NC contacts that came with their company's favorite model of relay. The NO and NC contacts from an internally generated PLC relay can be used an unlimited number of times.
10. If your circuit uses an internal PLC relay, and you use only one contact of that relay in the rest of the circuit, review the use of the relay. You should be able to revise the circuit and eliminate the internal PLC relay.
11. Timers can be a source of trouble if you unintentionally assign the same address to two different timers. A problem arises, but there is no way to predict what the problem will be. Suffice it to say that the first timer in the sequence will be correct, and the second will be incorrect. Depending on the circumstances, the second timer may never be activated, or it may be activated with the time value of the first timer.
12. A flashing pilot light on your control panel can be used to emphasize a special alarm or sequence in your process. It is a simple task to add one of the special internal timing relays to your circuit that

are provided in your PLC and thus achieve the continuous on-off flashing effect. (For more details, refer to the section entitled "Flashing Light Circuits" in chapter 5.)

TESTING

13. Purchase a spare PLC, preferably the same model that you are using in your plant operations. It will serve three purposes. First, it is a true spare part in your plant, ready to be installed in any type of emergency. Second, the spare PLC can be used as a program-testing tool. Keep it handy and have it prewired with power, three or four input push buttons, and a two-position or a three-position selector switch. It just makes good sense to take a section of a circuit that you have programmed and load it into the PLC to verify that the circuit is operating exactly the way you designed it. Third, the spare PLC can be used as a Programmable Read-only Memory (PROM) programmer. A PROM can be programmed in the office or lab and then be taken to the plant PLC or sent to an operation at one of your other plants.

INDEX